Advancing Sustainable Development Goals in Higher Education

Kaitano Dube · Liemiseng Masebala Tjabane
Editors

Advancing Sustainable Development Goals in Higher Education

Innovative Approaches for Teaching, Learning, and Student Success in South Africa

Springer

Editors
Kaitano Dube
College of Economic and Management Sciences
University of South Africa
Pretoria, South Africa

Liemiseng Masebala Tjabane
Centre for Academic Development
Vaal University of Technology
Vanderbijlpark, South Africa

ISBN 978-3-032-27645-2 ISBN 978-3-032-27646-9 (eBook)
https://doi.org/10.1007/978-3-032-27646-9

This Springer imprint is published by the registered company Springer Nature Switzerland AG
The registered company address is: Gewerbestrasse 11, 6330 Cham, Switzerland

Preface

This edited book examines the transformative role of higher education in advancing the Sustainable Development Goals (SDGs) from a South African perspective. *Advancing Sustainable Development Goals in Higher Education: Innovative Approaches for Teaching, Learning, and Student Success in South Africa* brings together transdisciplinary scholarship that explores curriculum transformation, inclusive pedagogy, digital innovation, student support systems, and sustainability-oriented practices. Through empirical case studies, theoretical reflections, and methodological diversity, the book showcases how universities respond to global challenges such as climate change, inequality, and technological disruptions while simultaneously addressing local developmental priorities. Emphasising innovation, equity, and relevance, this book positions higher-education institutions as critical agents of social change, knowledge production, and human capital development.

Pretoria, South Africa

Kaitano Dube
Liemiseng Masebala Tjabane

Acknowledgements

The editor would want to express sincere gratitude to Vaal University of Technology's Centre for Academic Development, the Faculty of Human Science and South Africa's Department of Higher Education and Training (DHET)'s University Capacity Development Programme (UCDP) for funding this book project. The editor would also want to thank the blind reviewers for their immense contribution to this project.

Contents

Editors and Contributors

About the Editors

Prof. Dr. Kaitano Dube is a Distinguished Professor at the University of South Africa, Pretoria in the College of Economic and Management Sciences. A former full professor of Tourism Geography at Vaal University of Technology and a visiting professor at Emirates Aviation University, Dubai, United Arab Emirates. Professor Dube is a leading research scholar in tourism, climate change, aviation, education, and sustainability. He is a leading research consultant and facilitator who has mentored several emerging scholars in the field. He holds several local and international university management positions. He is a C2 NRF Rated Researcher.

Liemiseng Masebala Tjabane is currently in Academic Staff Development in the Centre for Academic Development. Her role involves contributing to various teaching and learning development initiatives aimed at enhancing teaching and learning, as well as the scholarship on teaching and learning. She is also involved in teaching education studies at VUT at the undergraduate and postgraduate levels. Nationally, she acts as an advisor to the Teaching Advancement in Universities (TAU) project. She has researched and published in higher-education policy, educational development, teaching excellence, action research, and Recognition of Prior Learning (RPL). She is a member of the Higher Education Learning and Teaching Association of South Africa (HELTASA) and is currently participating in a project on Telagogy, focusing on Teaching and Learning in Society 5.0.

Contributors

Sean Bogle is an educator and practising photographer. He holds Magister Technologiae in photography from the Vaal University of Technology. Currently, he holds the position of lecturer in Diploma of Photography programme offered at Vaal University of Technology (VUT). His professional interests in research, teaching, and practice are the educational processes and applications used within the visual domain of photography to help communicate an idea, illustrate a concept, or tell a story. He uses both new and old technologies, such as digital and analogue photographic practices, to explore, practice, and reach a desired outcome within his visual context. He undertakes photographic commissions on a commercial basis, thus staying connected with current practices within the field of photography. He also explores photographic technologies and the intertwining web between old and new practices that help develop an independent voice and style for emerging student photographers.

Carmel Proctor has Ph.D. in Psychology from the University of Leicester, UK. She is Chartered Psychologist and Associate Fellow (AFBPsS) of the British Psychological Society (BPS) and the Director of PPRC Ltd. Carmel is co-author of Strengths Gym (2019), a character strengths-based educational course for children and adolescents. She has written and co-written multiple research papers and book chapters and has served on various editorial boards. Her interests are primarily in positive psychology, applied positive psychology, values, character, virtue ethics, personality, existential psychology, and well-being. Carmel is also Health and Care Professional Council (HCPC) registered Counselling Psychologist and Accredited Member of the British Association for Counselling and Psychotherapy (BACP). She works as a psychotherapist in Guernsey, UK.

Dr. Muntuwenkosi Chili is the executive director of the Centre for Academic Development at Vaal University of Technology. He holds a Ph.D. in Chemistry, awarded in 2011, and has built a distinguished career in higher education, spanning teaching, academic development, and institutional leadership. He began his academic career at the University of Zululand, serving as a lecturer and later as a senior lecturer from 2006 to 2013. In 2013, he joined Mangosuthu University of Technology as Deputy Director of the Teaching and Learning Development Centre, a role he held until 2021, where he contributed significantly to teaching and learning enhancement and institutional capacity building. Dr. Chili has strengthened his professional practice through Postgraduate Certificate in Education (UNISA), Management Development Programme at Stellenbosch University, and Postgraduate Diploma in Higher Education from Rhodes University. In 2023, he completed the Executive Development Programme at the University of the Free State and attained the TAU Fellowship status. He is currently pursuing an MBA in Leadership and Management at Haaga-Helia University of Applied Sciences in Finland.

Prof. Rabelani Dagada is Head of the Department of Digital Transformation and Supply Chain Management and Associate Professor of ICT at the UNISA Graduate School of Business Leadership. He previously worked at the Institute for Intelligent Systems (IIS) at the University of Johannesburg. Prior to joining the ISS, he served as Lecturer of Information Systems Management at the Wits Business School for just over 10 years. While at the Wits Business School, he introduced a programme on ICT Governance and Risk Management and served as its founding programme director. Prof. Dagada has published and edited many books. He holds Masters in Educational Computing (Rand Afrikaans University), Masters of Commerce in Information Systems (Wits), Ph.D. in Information Systems (UNISA), and Ph.D. in Developmental Economics.

Dr. Johannes Segone Dirane holds Doctorate in Education from Tshwane University of Technology. He is an experienced academic and higher education practitioner with expertise in research, staff evaluation and development, the Scholarship of Teaching and Learning (SoTL), risk and compliance, life skills, and educational technology. His technical competencies include Computer Applications Technology, A+ and N+ certification, HTML, and Dreamweaver. His research and professional interests focus on podcasting, e-learning and mobile learning (m-learning), cloud computing, ubiquitous learning, mobile and information technologies, technology-enhanced teaching and learning, and collaborative enquiry-based learning. Dr. JS Dirane has published widely in accredited academic journals, contributed chapters to scholarly books, and presented research papers at national and international conferences. Dr. JS Dirane currently serves as Staff Evaluation Manager at the Centre for Academic Development at the Vaal University of Technology, where he contributes to academic quality assurance, staff development, and institutional enhancement initiatives.

Rene van Eck is an associate professor known for her commitment to empathetic, student-centred teaching and impactful academic leadership. She earned her Ph.D. at 47 and became a professor at 52, exemplifying lifelong learning and intellectual resilience. Her academic journey is shaped by a belief in merit-based achievement and the meaningful support she received from key mentors along the way. Rene is particularly passionate about fostering engagement, connection, and holistic student development in the academic setting. Her work reflects a balance of scholarly excellence, compassion, and dedication to empowering others through education.

Dr. Sinikiwe Gwatidzo is currently Registrar at the Great Zimbabwe University. She holds Doctor of Business Administration from the Binary University of Management & Entrepreneurship, Malaysia. She is interested in Business and Strategic Management, Governance, Labour Market Dynamics, and Public Policy. She has lectured on Labour Economics and Change Management. As an administrator in higher education, she is passionate about the management of higher education institutions, quality in higher education, and evolving trends in education and training.

Dr. Anthony K. Isabirye is a distinguished South African scholar, educator, and human resource management specialist with over three decades of experience in the secondary, higher, and professional education sectors. He holds Doctor of Education in Curriculum Studies from the University of South Africa, a masters in human resource management, and a masters in education (Cum Laude) in Adult and Community Education from the University of Johannesburg, alongside earlier qualifications from Makerere University. From 2010 to 2022, Dr. Isabirye served as Senior Lecturer and Head of the Department of Human Resource Management at the Vaal University of Technology, where he provided academic leadership, supervised postgraduate research, and contributed extensively to curriculum development and quality assurance. He previously worked as Assistant Researcher at UNISA and as Education Specialist in Mpumalanga Province. Dr. Isabirye is registered with the South African Council for Educators (SACE), South African Board for People Practices (SABPP), and ETDP-SETA. His research interests include human resource practices, organisational justice, professional development, innovative teaching and learning, employee engagement, and public sector performance. He has published widely in accredited journals and continues to supervise the postgraduate students. Currently, he is active in research, part-time teaching, and serves as a director at Quantum Executive Education (Pty) Ltd., focusing on human resource development and curriculum design.

Lerato Jerome Mosebi is Communication Skills/Education Lecturer at VUT and an aspiring Hip Hop Pedagogy researcher. He is passionate about strategically making poetry in South African schools inclined to the local poetry in music that learners naturally love, instead of poetry from the West, such as Shakespeare. Lava is also an amateur golfer and tennis player who writes hip-hop music for pleasure during his spare time.

Hendrick Musawenkosi Langa obtained B.Sc. degree in Computer Science and Information Systems from the University of South Africa, B.Tech degree in Electrical Engineering from the Vaal University of Technology, and both M.Phil and D.Phil degrees in Electrical Engineering from the University of Johannesburg. He also completed his Master of Business Leadership (MBL) degree at the Unisa School of Business Leadership. His research interests range from Renewable Energy Systems, Matrix Converters and Power Electronics, Artificial Intelligence Applications, Web Computing and Mobile App Development, to Engineering Education. He has published more than three book chapters, five journal articles, and over 25 conference papers. He has hosted three post-doctoral fellows and published over seven accredited articles. He supervises M and D students in the School of Computing at the University of South Africa, where he is currently an associate professor of Computer Vision in the Centre for Augmented Intelligence and Data Science. Prof. Dr. Langa is a senior member of the South African Institute of Electrical Engineers and a member of the Institute of Electrical Electronic Engineers.

Dr. Mashaole Jacob Makwela is an artist, animator, filmmaker, educator, and researcher. He currently serves as a senior lecturer and course co-ordinator for the Fine Art programme in the Department of Visual Arts and Design at the Vaal University of Technology. His research interests include animation studies, liberation struggle history, film studies, new media, AIKS, visual studies, art and technology, digital arts education, and entrepreneurship from an African-centred perspective. His teaching responsibilities include digital imaging, storytelling, animation, visual design, video, film, art theory, and visual studies. Additionally, he is involved in curriculum development for the programme. His community engagement activities include training, skills transfer, mentoring, and guiding the generation of ideas for visual creative projects. His creative projects focus on moving images, such as animation, film, music videos, promotional videos, and cartoons.

Prof. Lazarus Maleho is the executive dean of the Faculty of Human Sciences, Vaal University of Technology, South Africa. The faculty he is heading comprises of four departments, Department of Education, Legal Sciences, Tourism and Integrated Communication, and Visual Arts and Design.

Prof. Lazarus has more than 15 years' experience in academia, among the duties and roles he played include lecturing, programme co-ordination, and head of department. Lazarus played a vital role in curriculum development of diplomas and postgraduate diplomas. He was also instrumental as a chair of the task team for assessment policy at VUT. Lazarus was involved in many policy developments at VUT and he seats in number of committees. He successfully supervised and co-supervised masters and Ph.D. students. He is currently supervising five masters' students. He mentored more than 20 B.Tech students. He received faculty research award in 2017, first place research achievement in the faculty. Lazarus was also acknowledged for contributing towards research in 2016. Lazarus attended and presented papers at a number of conferences both national and international. He published papers in both accredited and non-accredited journals. His background is in strategic communication, public relations, and media studies.

Dr. Luyanda Marhaya is an experienced higher education leader with strong expertise in academic planning, quality assurance, and institutional governance. He currently serves as Director at the Vaal University of Technology, where he drives strategic initiatives that enhance academic quality, policy coherence, and curriculum development in support of the university's academic mission. Previously, Dr. Marhaya was Director of Academic Planning and Quality Assurance at the University of the Free State, providing strategic leadership in programme development, institutional reviews, and accreditation to ensure alignment with national and international quality standards. He also served as Senior Manager at the Council on Higher Education (CHE), South Africa's statutory quality assurance body.

Dr. Marhaya holds Postgraduate Diploma in Higher Education from Rhodes University, Masters in Higher Education from the University of the Free State, and Ph.D. in Education with a focus on Higher Education. He is Reference Group Member at the Department of Higher Education and Training, appointed by the minister, and a panel reviewer for the CHE's sector-wide review on online and blended learning. He is also a member of SAAIR and HELTASA.

Dr. Munienge Mbodila is a distinguished academic and researcher who serves as Manager, Student Tracking and Institutional Research, and Senior Lecturer in the Department of Networks and IT Support in the Faculty of Engineering, Built Environment and IT. As a leading expert in higher education, Dr. Mbodila is dedicated to advancing student success and institutional decision-making through data-driven strategies. He has made significant contributions to the academic community. Overall, he has 16 years of teaching experience in HE, 13 years of research experience, and over 70 published articles, as well as numerous presentations at national and international conferences. His goal as a lecturer/teacher is to develop students' interest in ICTs and to encourage them to familiarise themselves with modern computer technologies and their productive use in their education, lives, and professional development. His dedication to advancing student success through data-driven decision-making turns institutional data from a static record of the past into a dynamic blueprint for building successful student success initiatives and an equitable future for every learner.

Malcolm Phafane is a South African photographer and a lecturer within the Department of Visual Arts and Design at the Vaal University of Technology (VUT). He recently completed his Master of Fine Arts (MAFA) at Witwatersrand University. His achievements include documenting the resilience of his community in Sharpeville. Balancing his professional practice with academia, Phafane specialises in teaching core technical and conceptual modules which include Media Practice, Applied Media Aesthetics, and Theory of Media Technology. His pedagogy is deeply informed by his personal expertise in social concern/documentary photography, where he encourages students to master traditional photographic aesthetics while engaging with their local communities. By bridging the gap between classroom theory and field-based documentation, Phafane prepares his students to use the camera as both a professional tool and a medium for social advocacy and upliftment.

Zanele Lucia Radebe is a lecturer and a first year co-ordinator in Fine Art, at the Vaal University of Technology, South Africa. She holds Master of Technology (M.Tech) in Fine Art and a Postgraduate Diploma in Higher Education (PGDHE). Her research is grounded in gender studies (mainly African women in the visual arts) and art education. She is interested in documenting the lives and works of African Visual Artists. Radebe is also passionate about integrating theory and practice in the classroom. Her work contributes meaningfully to the broader discourse on gender, cultural identity, and art education.

Corne Venter is a Johannesburg-based educator and practising artist currently affiliated with the Department of Visual Arts and Design at the Vaal University of Technology. He holds a masters degree in visual arts from the University of Johannesburg, a qualification that underpins his dual commitment to rigorous academic research and creative work. His professional practice explores the complex themes of memory and trauma, with a specific focus on pursuing resolution and discovering meaning within the personal narrative. His work is found in both academic and private galleries and is represented in several private collections. This creative output informs his contribution to the current collaboration, where he draws on his expertise in visual storytelling and the creative process's restorative potential. Through both his teaching and studio practice, Venter investigates the intersections of formal aesthetic enquiry and the human capacity for resilience, seeking to transform the fragments of the past into a coherent understanding.

Part I
Background and Introduction

Chapter 1
Advancing Sustainable Development Goals in Higher Education

Kaitano Dube, Masebala Liemiseng Tjabane, and Muntuwenkosi Chili

Abstract Globally, higher education plays a central role in the advancement of the Sustainable Development Goals (SDGs) through teaching, learning, research community engagement, engaged scholarship, and internationalisation. This edited book addresses how higher education is evolving its curriculum to deal with increasing global risks such as climate change, environmental degradation, technological disruptions, and increasing socioeconomic and political challenges. This 13-chapter edited book utilises qualitative and quantitative research methodologies in various sections. Across various chapters, the book demonstrates the role of South African universities in ensuring the country's sustainability agendas. Various chapters demonstrate that universities in South Africa play a critical role in the quest for sustainability as sites of knowledge production and critical agents for socioeconomic transformation, utilising innovative and inclusive techniques. This chapter provides an overview of how the education system in South Africa is addressing contemporary geospatial, environmental, and technological challenges. The chapter underscores the need for higher education institutions to continue being responsive and develop curricula that are transformative and future-oriented. The book positions South Africa's higher education at the core of ensuring sustainable development.

Keywords Digital transformation · Social inclusion · Teaching and learning innovation · Global risks · Student success · Curriculum transformation

K. Dube (✉)
College of Economic and Management Sciences, University of South Africa, Pretoria, South Africa
e-mail: dubekaitano@gmail.com

Emirates Aviation University, Dubai Academic City, Dubai, UAE

M. L. Tjabane
Centre of Academic Development, Vaal University of Technology, Vanderbijlpark, South Africa
e-mail: masebala@vut.ac.za

M. Chili
Centre for Academic Development, Vanderbijlparl, South Africa

K. Dube and L. M. Tjabane (eds.), *Advancing Sustainable Development Goals in Higher Education*, https://doi.org/10.1007/978-3-032-27646-9_1

1.1 Background and Introduction

The adoption of the Sustainable Development Goals (SDGs) in 2015 was a historical moment in human development, as it allowed humanity to reimagine a common future anchored on a unified vision. The adoption of the 17 SDGs ushered in a new era of conservation, environmental protection, and, in many respects, an economic development trajectory that considers the need to balance the socioeconomic and environmental demands of society. The adoption of SDGs calls for society to ensure social and environmental justice in many respects. The 249 targets that emanated from the Agenda for Sustainable Development were geared towards ensuring that as the world develops, the cares of present and future generations' quest for a global order are well articulated and represented.

At the core of sustainable development is the realisation of the threats posed by pollution, environmental degradation, climate change and other global threats (Li et al., 2025; Zhu et al., 2022). The SDGs are about creating a transformative developmental agenda which captures the essence of sustainable development in the midst of a transformed environment and reversing the damage caused by pollution and climate change (Andersen, 2024). In many respects, the SDGs are about being responsive to threats to the global order. Sustainability is about being responsive to all forms of challenges that have been flighted by the World Economic Forum in the last 10 years (2015–2025). Sustainability is about the world being responsive to many of the challenges identified in the 2026 Global Risks Report, as shown in Fig. 1.1.

These risks call for a properly coordinated approach to resolving global challenges, as they span multiple disciplines. The top ten risks for 2026 show that the world must take urgent steps to address environmental challenges such as extreme weather events and pollution. In the upcoming years, the world must address several environmental problems which can threaten sustainability, such as extreme weather events, biodiversity loss, changes in earth systems, shortage of natural resources, and pollution, under the leadership of capable and resourced higher education institutions.

However, the world must grapple with a range of societal issues, such as geopolitical issues with geoeconomic confrontation and state-based conflict, amidst the rise of a multipolar world with all its complexities and challenges (Stephen, 2025). The emergence of these geopolitical issues is concerning as they stand to redefine the world order at a time when the world is grappling with increasing environmental challenges which seem complex and financially costly. The rise of extreme nationalism (Bieber, 2018), xenophobia (Agaigbe & Akuraga, 2025; Kersting, 2009), and non-adherence to the global rules order presents challenges to the achievement of the SDGs. The movement from a unipolar world to a multipolar world presents unique challenges and opportunities for current and future societal and economic development trajectories.

From a societal perspective, the geopolitical and environmental challenges highlighted above increase social polarisation, deepen inequality, threaten the respect for human rights, and in many ways create a chaotic world where development cannot be

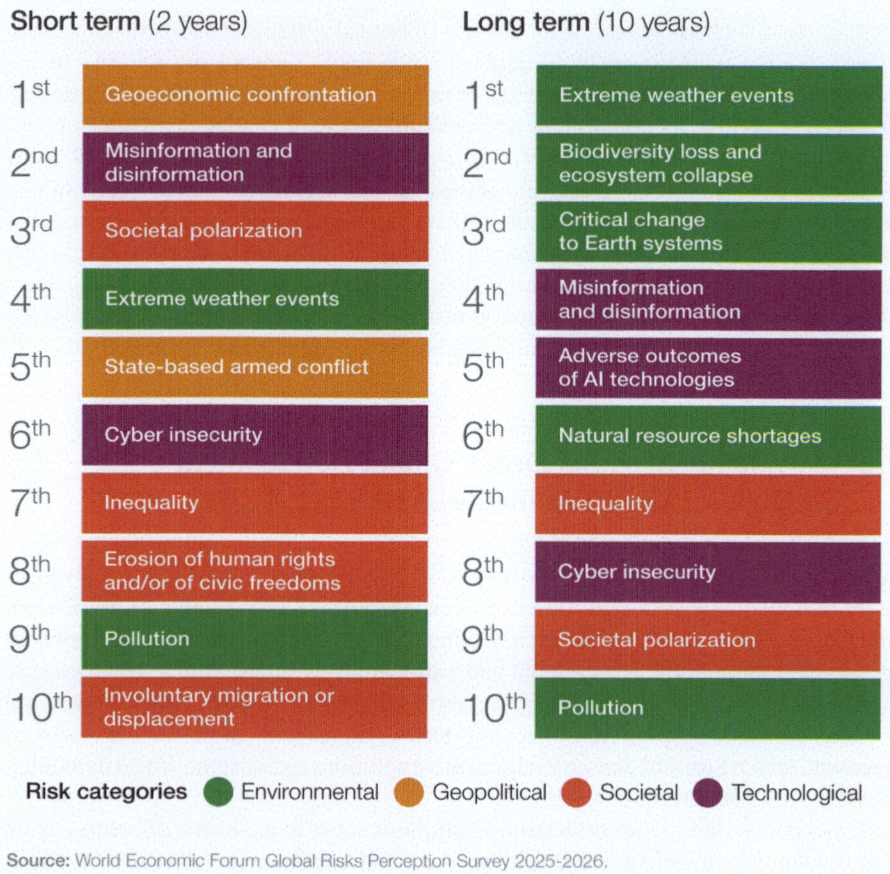

Fig. 1.1 2025–2026 top ten global risks *Source* World Economic Forum (2026)

sustained. With the rise of extreme nationalism (Hazel, 2025) and the advancement of individualistic narratives, there is a threat to societal stability, triggering forced migration (Li et al., 2025; Lundström & Poletti Lundström, 2025) and disrupting social cohesion.

The role of technology has come under scrutiny, and the threats that technology poses both in fuelling societal instability and in the erosion of the social fabric (Cicek et al., 2025; Li et al., 2025; Viale Pereira et al., 2025). Social media has been used to create all sorts of misinformation (Wang et al., 2025), building falsehoods and fomenting rhetoric and violence that creates social instability, whether against migrants or in spreading extremely racist ideologies (Kolluri et al., 2025). Social media and electronic platforms have also been used to spread misinformation and deceptive messages about climate change (Kresin et al., 2025).

The growth of the Internet of Things presents many opportunities for the world, but in almost equal measure is the threat that arises from the increased use of computer technologies, such as cybersecurity threats (Dube et al., 2025a, 2025b). Sustainability means that the world must build adequate measures to ensure the security of the computing public. The rise of AI technology post the COVID-19 pandemic has been celebrated for its potential to address some societal challenges (Galetsi et al., 2024; Ghazi et al., 2025); however, it brings its own set of challenges which must be addressed to ensure sustainability across various sectors of the economy. In this regard, higher education institutions (HEIs) have their work cut out in terms of what needs to be addressed. Sustainable Development cannot be achieved without a robust and engaged mass scholarship aimed at addressing some of these challenges. Universities and other higher education institutions have an obligation to assist the world in navigating these aspects in a manner that ensures sustainability.

1.2 Advancing SDGs to Foster Global Development in Higher Education Institutions

A closer look at global risks demonstrates the areas which require skills and advancement to holistically deal with threats to sustainability. Various fields of curriculum and responsive curriculum, research, technology, and innovation must be employed to ensure sustainability at the global and national levels (Dube et al., 2024b). There will be a demand for earth system sciences, social scientists, political scientists, and technology-based responsive curricula (Dube et al., 2024a) to be the key focus of universities to foster and champion innovative solutions to assist the world in moving forward with the current challenges.

Universities and higher education institutions must make every effort to ensure that students are capacitated with relevant Artificial Intelligence (AI) (Chere & Wayi-Mgwebi, 2024; Mudhombo, 2024), computer, data, and cybersecurity skills across various disciplines of learning to better prepare students for the world they are living in and to be better fit employees and entrepreneurs that can fit into the world of tomorrow. Given the demand for environmental protection, the curriculum must be revised to ensure that every student is equipped with sustainability and green technology skills as a streamlined intervention to create an environmentally conscious citizen.

The shift in geopolitics and geoeconomics alongside social changes calls for higher education institutions to produce candidates with high emotional intelligence and agile leaders with sound judgement who can adapt to ensure that they can perform to solve complex problems which threaten sustainability.

1.3 Space for SDG Advancement in South African Higher Education Space

Higher Education Institutions play a central role in the advancement of SDGs and, in many ways, act as a catalyst for development and change in society (Dube et al., 2024c). Universities assist the SDG agenda through transformative research, innovation, and community engagement. Countries have learned to use higher education institutions to address their SDG challenges and foster a better world.

Higher education institutions in South Africa have been working hard to ensure curriculum SDG alignment through the revision of pedagogical approaches (Dube et al., 2024a). Another key focus of HEIs in South Africa has been to ensure the expansion of access to higher education in a manner that addresses historical imbalances and exclusions (Chankseliani & McCowan, 2021). The objective of South African higher education is to ensure the inclusion of previously disadvantaged communities, such as women and Black people. Higher education institutions have also been making frantic efforts to ensure the promotion of lifelong learning in a manner that embraces all. Higher education institutions in South Africa are also embracing digital technology and championing digital literacy to ensure that society has the requisite technical skills to ensure the delivery of good quality of life (Nhamo et al., 2024).

Various institutions and campuses have different approaches to SDG localisation for quality education delivery. HEI embraces various sustainability initiatives aimed at addressing local and global problems. These initiatives range from those aimed at addressing environmental sustainability to those aimed at ensuring social equity in the supply chain. These projects are delivered through various foras, such as community engagement and engaged scholarship. Joint research and innovation projects are undertaken by various institutions to foster community development and enhance education delivery.

Addressing SDGs in higher education institutions has its own set of challenges (Mawonde & Togo, 2021; Nhamo, 2021). A key challenge in some institutions is the lack of a thorough understanding of the SDGs by university personnel. In addition, due to increased budget cuts, some institutions have been struggling to commit and invest in the SDGs and capacitate SDG initiatives. Institutional instability has also been flagged as a real challenge in ensuring that all higher education institutions are on track to address SDGs. Mawonde and Togo (2019) argued that some universities in South Africa have established administrative structures dedicated to sustainability to ensure global campus adoption of sustainability using sustainability champions.

This book aims to explore some of the ways in which universities in South Africa are catching up to the sustainability agenda under the new global order to ensure relevance and responsiveness to the broader issues of sustainability. The next section outlines the structure and principal contents of this book.

1.4 Book Overview, Structure and Contents

The higher education system is undergoing a significant transformation due to various factors. This transformation is necessitated by the pressing need to address various challenges that are being faced by the world, including poverty, inequality, environmental degradation, and climate change, with a view to ensuring that the world achieves its quest for sustainable development. The South African journey to sustainability is being shaped by the country's rich cultural heritage and complex socio-cultural history. The country's unique context calls for innovative approaches to sustainable development that are globally relevant and locally appropriate. The approach must be grounded in the local context. It is from this background that the book Advancing Sustainable Development Goals in Higher Education: Innovative Approaches for Teaching, Learning and Student Success in South Africa was birthed. This edited book presents a diverse range of scholarly perspectives which explore institutions that can foster inclusive, effective, and sustainable educational environments.

This book builds upon the first volume of a similar work, Redefining Education and Development: Innovative Approaches in the Era of Sustainable Goals (Dube, 2024b). This book delves into the critical roles of educators and institutions. The book is divided into four sections, with the second section covering three chapters that cover the philosophical and sociocultural foundations of education. This section deals with inclusivity, cultural relevance, and reflective teaching practice. Chapter 2 which is the first chapter of this section, examines cultural influences on peer group evaluation and the role of Ubuntu and shared intentionality in higher education. The second chapter of this section delves into reflective teaching through character strength, making use of a case study of the TAU programme. The third chapter of that section, which is Chapter 5 of the book, looks at the significance of using hip hop pedagogy in teaching poetry in the Gauteng Province of South Africa.

The third section of the book focuses on student success, inclusion, and support systems. This section focuses on ensuring equitable access and monitoring of long-term outcomes of the higher education system in South Africa. The section opens with a chapter that focuses on the implementation of extended curriculum programmes in rural universities in South Africa. Chapter 6 proposes a multidimensional protocol for identifying at-risk university students. Chapter 7 assesses students' success through alumni achievements in visual arts and design.

The fourth section of the book is on digital transformation and technological infrastructure which focuses on the need for technology to achieve the aspirations of SDG 4. Chapter 8 is the first chapter in this section which deals with information and communication technology education for enhanced teaching and learning. Chapter 9 focuses on the design and implementation of a smart client-server attendance register to minimise latency and enhance flexibility.

The fifth section focuses on sustainability and a global perspective. This section focuses on education for sustainable development and global citizenship (GC). Chapter 10 deals with the role of higher education in the green technology revolution, while Chap. 11 tackles the issue of practice-based exploration of camera obscura and smartphone photography to achieve economic sustainability. Chapter 12 deals with how the internationalisation of universities can be used to enhance research impact and productivity.

1.5 Methodological Approaches

This book utilises a diverse and interdisciplinary approach which prioritises qualitative and interpretive frameworks to investigate the lived experiences of students and lecturers in South Africa. Several chapters make use of qualitative research designs which range from case studies, interviews, and focus groups to examine social phenomena such as the implementation of extended curriculum programmes (Chap. 5) and the integration of Hip Hop Pedagogy in poetry (Chap. 4). In other chapters, the authors utilise a practitioner-based approach. This approach is also used in Chap. 11, where it is utilised to investigate pedagogical growth and innovative artistic techniques such as camera obscura. This approach allows for a deep and thematic exploration of how cultural philosophies such as Ubuntu intersect with academic assessment and student success.

The book also utilises quantitative and systematic review methodologies to provide a broader evidence-based approach to institutional improvement. For example, Chap. 8 utilises quantitative analysis of more than 1900 research participants to assess the impact of technology-enhanced learning on pass rates, whereas Chap. 9 utilises software engineering approaches to design and test a smart client–server system for attendance. Chapter 6 uses a Prisma-guided systematic literature review to develop student risk protocols and bibliometric analysis in Chap. 10 to map the global role of universities in the green technology revolution. Through the blending of interpretive storytelling with data-driven analysis, this edited book offers a holistic methodological toolkit for addressing the Sustainable Development Goals within the higher education sector.

1.6 Conclusion

The book covers various chapters and topics that are central to SDG 4. Each chapter responds to the various targets of SDG 4. It can be argued that the book is conceptualised around at least four broader thematic focus areas, namely, chapters that focus on equitable access and inclusion which resonate with SDG targets 41 and 4.5. The relevant chapters focus on those aimed at ensuring the removal of barriers for marginalised or underrepresented students. The other section of the book deals with skills for the future and digital literacy which resonate with target 4.4. These chapters highlight the technical and vocational skills required for employment. The third batch of chapters, aligned to target 4a and c, is aimed at addressing and enhancing the human and infrastructure elements of classrooms. The last batch of chapters is aligned with SDG target 4.7, dealing with long-term outcomes and alumni success. This is critical because education for sustainable development is measured by the impact graduates have on the world.

References

Agaigbe, F. M., & Akuraga, F. (2025). Globalisation, migration and xenophobia: An investigation of the issues, challenges and the way forward. *Kashere Journal of Politics and International Relations, 3*(2), 404–417. https://journals.fukashere.edu.ng/index.php/kjpir/article/view/530

Andersen, T. J. (Ed.). (2024). Prelims. In *Sustainable and resilient global practices: Advances in responsiveness and adaptation* (p. 0). Emerald Publishing Limited. https://doi.org/10.1108/978-1-83797-611-920241010

Bieber, F. (2018). Is Nationalism on the Rise? Assessing Global Trends. *Ethnopolitics, 17*(5), 519–540. https://doi.org/10.1080/17449057.2018.1532633

Chankseliani, M., & McCowan, T. (2021). Higher education and the sustainable development goals. *Higher Education, 81*(1), 1–8. https://doi.org/10.1007/s10734-020-00652-w

Chere, M., & Wayi-Mgwebi, N. (2024). Integration of generative artificial intelligence in higher education: Pedagogy factors and best practices. In K. Dube (Ed.), *Redefining education and development: Innovative approaches in the era of the sustainable development goals* (pp. 93–112). Springer Nature Switzerland. https://doi.org/10.1007/978-3-031-69954-2_7

Cicek, M., Gursoy, D., & Lu, L. (2025). Adverse impacts of revealing the presence of "Artificial Intelligence (AI)" technology in product and service descriptions on purchase intentions: The mediating role of emotional trust and the moderating role of perceived risk. *Journal of Hospitality Marketing & Management, 34*(1), 1–23. https://doi.org/10.1080/19368623.2024.2368040

Dube, K. (Ed.). (2024b). *Redefining education and development: Innovative Approaches in the era of the sustainable development goals*. Springer Nature Switzerland. https://doi.org/10.1007/978-3-031-69954-2

Dube, K., Booysen, R. M., & Chili, M. (2024a). Education issues post-COVID-19: Advancing sustainability—A policy perspective. In K. Dube (Ed.), *Redefining education and development: Innovative approaches in the era of the sustainable development goals* (pp. 199–208). Springer Nature Switzerland. https://doi.org/10.1007/978-3-031-69954-2_13

Dube, K., Booysen, R. M., & Chili, M. (2024b). Redefining education and development: Innovative approaches in the era of sustainable goals. In K. Dube (Ed.), *Redefining education and development: Innovative approaches in the era of the sustainable development goals* (pp. 1–8). Springer Nature Switzerland. https://doi.org/10.1007/978-3-031-69954-2_1

Dube, K., Booysen, R. M., & Chili, M. (2024c). Redefining education and development: Innovative approaches in the era of sustainable goals. In K. Dube (Ed.), *Redefining education and development: Innovative approaches in the era of the sustainable development goals* (pp. 1–8). Springer Nature Switzerland. https://doi.org/10.1007/978-3-031-69954-2_1

Dube, K., Gwatidzo, S., & Maleho, L. (2025a). Opportunities and challenges of AI adoption in higher education, a post COVID-19 analysis. In *2025 3rd International Conference on Business Analytics for Technology and Security (ICBATS)*, 1–5. https://doi.org/10.1109/ICBATS66542.2025.11258206

Dube, K., Sixolisile, & Mdoda, S. (2025b). Cyber security, an evaluation of key issues in the airline industry. In *2025 3rd international conference on cyber resilience (ICCR)*, 1–5. https://doi.org/10.1109/ICCR67387.2025.11292034

Galetsi, P., Katsaliaki, K., & Kumar, S. (2024). Realizing resilient global market opportunities and societal benefits through innovative digital technologies in the post COVID-19 era: A conceptual framework and critical literature review. *IEEE Transactions on Engineering Management, 71*, 10650–10666. https://doi.org/10.1109/TEM.2023.3303080

Ghazi, K., Kattara, H., Salem, I. E., & Shaaban, M. N. (2025). Benefit-triggered or trust-guided? Investigation of customers' perceptions towards AI-adopting hotels amid and post COVID-19 pandemic. *Tourism and Hospitality Research, 25*(1), 38–56. https://doi.org/10.1177/14673584231184161

Hazel, M. (2025). The madness behind the method: What's really beneath anti-critical race theory legislation. *Educational Policy, 39*(7), 1357–1391. https://doi.org/10.1177/08959048241305527

Kersting, N. (2009). New nationalism and xenophobia in Africa—A new inclination? *Africa Spectrum, 44*(1), 7–18. https://doi.org/10.1177/000203970904400102

Kolluri, A., Murthy, D., & Vinton, K. (2025). Quantifying the spread of racist content on fringe social media: A case study of Parler. *Big Data & Society, 12*(2), 20539517251321750. https://doi.org/10.1177/20539517251321752

Kresin, S., Kremer, K., Nehring, A., & Büssing, A. G. (2025). Students' awareness and conceptions of science-related communication mechanisms on social media. *Journal of Research in Science Teaching, 62*(3), 756–791. https://doi.org/10.1002/tea.21973

Li, C., Zhang, X., & Wang, L. (2025). How industry–university–research integration promotes green technology innovation in Chinese enterprises: The dual mediating pathways and nonlinear effects. *Systems, 13*(8), 696. https://doi.org/10.3390/systems13080696

Lundström, M., & Poletti Lundström, T. (2025). Radical nationalism. *Journal of Political Ideologies, 30*(1), 84–97. https://doi.org/10.1080/13569317.2023.2241384

Mawonde, A., & Togo, M. (2019). Implementation of SDGs at the University of South Africa. *International Journal of Sustainability in Higher Education, 20*(5), 932–950. https://doi.org/10.1108/IJSHE-04-2019-0156

Mawonde, A., & Togo, M. (2021). Challenges of involving students in campus SDGs-related practices in an ODeL context: The case of the University of South Africa (Unisa). *International Journal of Sustainability in Higher Education, 22*(7), 1487–1502. https://doi.org/10.1108/IJSHE-05-2020-0160

Mudhombo, I. (2024). The emergency of chatbots and student learning dilemma: A comprehensive review. In K. Dube (Ed.), *Redefining education and development: Innovative approaches in the era of the sustainable development goals* (pp. 151–160). Springer Nature Switzerland. https://doi.org/10.1007/978-3-031-69954-2_10

Nhamo, G. (2021). Localisation of SDGs in higher education: Unisa's whole institution, all goals and entire sector approach. *Southern African Journal of Environmental Education*. https://doi.org/10.4314/sajee.v37i1.5

Nhamo, G., Chapungu, L., & Dube, K. (2024). Universities and sustainable development goals localisation: Insights from academic staff in Zimbabwe. *Cogent Education, 11*(1), 2428877. https://doi.org/10.1080/2331186X.2024.2428877

Stephen, M. D. (2025). The diffusion of global power and the decline of global governance | ethics & international affairs | Cambridge core. *Ethics & International Affairs*. https://doi.org/10.1017/S0892679425100129

Viale Pereira, G., Ciesielska, M., Rizun, N., & Alexopoulos, C. (2025). Exploring adverse effects from technological change in the context of smart city governance. *Journal of the Knowledge Economy*. https://doi.org/10.1007/s13132-025-02875-7

Wang, J., Zhai, Y., & Shahzad, F. (2025). Mapping the terrain of social media misinformation: A scientometric exploration of global research. *Acta Psychologica, 252*, Article 104691. https://doi.org/10.1016/j.actpsy.2025.104691

World Economic Forum. (2026, January 14). *Global risks report 2026: Top risks in an 'age of disorder'*. World Economic Forum. https://www.weforum.org/stories/2026/01/global-risks-2026-top-10-two-and-ten-year-horizon/

Zhu, J., Zhai, Y., Feng, S., Tan, Y., & Wei, W. (2022). Trade-offs and synergies among air-pollution-related SDGs as well as interactions between air-pollution-related SDGs and other SDGs. *Journal of Cleaner Production, 331*, Article 129890. https://doi.org/10.1016/j.jclepro.2021.129890

Part II
Philosophical and Sociocultural Foundations of Education

Chapter 2
Cultural Influences on Peer Group Evaluation: The Role of Ubuntu and Shared Intentionality in Higher Education

Rene van Eck

Abstract Peer group evaluation serves as an effective mechanism for gauging individual performance and promoting collective development in collaborative educational and professional environments. This study delves into the factors that can influence the accuracy and effectiveness of peer assessments, focusing on the potential roles of the ancient African philosophy of Ubuntu, which underscores interconnectedness, empathy, and communal well-being, and the theory of shared intentionality, highlighting collaborative engagement with common objectives. A study was conducted with IT students in a Business Analysis course, where they evaluated their group members' participation in an assignment. Concurrently, individual, confidential, semi-structured interviews were conducted to explore students' perceptions of their peers' contributions. The central aim was to identify factors contributing to instances where peer evaluations do not align with expected outcomes, an area that has been less explored in the current literature. Utilising a mixed-methods approach that combines peer evaluation data with thematic analysis of interview insights, this study seeks to illuminate the intricate dynamics at play in peer assessment scenarios, particularly how a strong emphasis on group cohesion can lead to potentially inflated evaluations and a reluctance to provide truthful criticism regarding individual contributions. A total of 48 students participated in the peer evaluations, representing the entire population, and 21 students participated in the interviews. Initial findings indicate that while Ubuntu fosters a supportive atmosphere, it might inadvertently discourage candid and critical feedback to avoid negatively affecting peers. The presence of shared intentionality also appears to affect the honesty of student evaluations. This study highlights the tension between cultural values, such as Ubuntu, and the need for honest feedback, offering insights for designing culturally responsive peer assessment frameworks.

Keywords Peer group evaluation · Collaborative learning · Cultural bias · Assessment fairness · Ubuntu (philosophy) · Shared intentionality

---

R. van Eck (✉)
Department of Computer Sciences, Vaal University of Technology, Vanderbijlpark, South Africa
e-mail: rene@vut.ac.za

K. Dube and L. M. Tjabane (eds.), *Advancing Sustainable Development Goals in Higher Education*, https://doi.org/10.1007/978-3-032-27646-9_2

2.1 Introduction

In the context of higher education, it is essential for educators to move beyond the sole development of academic skills, such as analytical reasoning and problem-solving, and to intentionally integrate opportunities for cultivating interpersonal and professional competencies within the classroom (Barac et al., 2021; Rajabzadeh et al., 2022). This holistic approach is critical for preparing students to contribute effectively to and thrive in the future workforce. (Shah-Fairbank et al., 2021). Collaborative learning refers to educational situations in which students work together in small groups to enhance their understanding of the work they are engaging with (Babo et al., 2021; Freeman, 1995).

Group work has become an increasingly important aspect of higher education, offering students opportunities to engage in projects that mirror workplace scenarios, foster teamwork, and develop essential communication and collaboration skills (Casanova, 2022; Cheng & Warren, 2000; Mahat et al., 2024). Peer assessment, or peer review that goes hand in hand with peer group work, is a process in which individuals within a similar status or group evaluate the quality of their peers' work, performance, or contributions based on predefined criteria (Gaynor, 2020; Lejk & Wyvill, 2002; Stenberg et al., 2021). A study in South Africa indicated that participants believed that peer learning could facilitate social, cognitive, behavioural, collaborative, and social engagement and enhance academic development (Steenkamp & Brink, 2024). Although groupwork and peer assessment have some definite advantages, some scholars emphasise their negative aspects, such as students not being assessed according to their contributions to work development (Ashenafi, 2017; Babo et al., 2021; Kennedy, 2005).

Hall and Buzwell (2013) noted that the growing prevalence of group work in higher education has unfortunately coincided with more frequent instances of students not contributing equally. This problem can lead to negative experiences for other students within the group, thereby diminishing the benefits of collaborative learning. The significance of collaboration and teamwork in the dynamic world of education and professional settings cannot be overemphasised (Driskell et al., 2018; Tumpa et al., 2022). Student success relies heavily on engagement, and active learning methods can foster this engagement (De Castro et al., 2021; Hernández González & Blackford, 2022; Hu et al., 2024).

Strong evaluation systems are essential when people work in groups to accomplish shared goals (Linn et al., 2022). As a fundamental component of collaborative settings, peer group evaluation provides a special perspective that allows people to obtain feedback, improve their abilities, and support the progress of the group as a whole (Barac et al., 2021). Cheng and Warren (2000) developed and proposed procedures to factor in the contributions of individual group members engaged in an integrated group project using peer assessment procedures. Peer group evaluation is a valuable tool for assessing individual contributions within a group and reflecting the dynamics of the team as a whole. It works under the assumption that peers are in a good position to offer helpful criticism because they have first-hand knowledge of their colleagues' work. This procedure goes beyond compiling individual work; it can also develop a culture of accountability and mutual support, as well as a comprehensive understanding of group dynamics (Van Gennip et al., 2009).

The study described in this paper was guided by the following research questions:

1. What were the outcomes of the peer group assessment process in the context of group assignments?
   and
2. What factors, if any, influenced the effectiveness of peer group assessment?

This study aligns with Sustainable Development Goal 4 (SDG 4), which aims to ensure inclusive and equitable quality education and promote lifelong learning opportunities for all individuals.

Students were asked to evaluate the level of participation in the peer groups, and it was observed that without fail, in all the groups, each student scored their peers in the group to have had a 100% participation in the assignment. Through informal interviews with the students, they admitted that there were some students who did not participate at all, yet they scored them with a 100% participation on the score sheet.

In a study by Kench et al. (2009), the same trend was found, that a low percentage of students in the groups were identified as not making an acceptable contribution to the group project. Kench et al. (2009) argued that this may be due to either a high level of satisfaction with their peers' contribution to group work or a low level of willingness to report the poor performance of their peers. Watson et al. (2010) identified basic demographics, personality factors, exam scores, and observations of team process and team goal attainment as factors that may influence the effectiveness of peer group assessments. De Meulemeester et al. (2022) state the importance of keeping the teacher as an evaluator, since students' scores (peer evaluations) are not always reliable. Further investigation is warranted to determine why students behave in this manner.

While collaborative learning offers benefits, cultural norms such as Ubuntu and shared intentionality may uniquely shape peer assessment behaviours in non-Western contexts (Sefotho & Letseka, 2024). Despite this philosophical alignment, some educational environments remain competitive and individualistic, undermining Ubuntu's collaborative ideals. Ubuntu's emphasis on communal success can present challenges in contexts where individual contributions must be assessed and where it is essential to ensure that all group members actively participate (Akabor & Phasha, 2025).

Ubuntu is a deeply rooted traditional African idea centred on the principles of community service and is fundamentally defined by the understanding that an individual's existence is intrinsically linked to their active engagement within their culture and broader society (Chakabwata & Mukazi, 2022a). More than just a social construct, Ubuntu represents an African humanist concept that employs the powerful metaphors of 'brotherhood' or 'sisterhood' to forge a profound connection between the individual and the encompassing world, thereby fostering a strong sense of collective identity and shared purpose. Furthermore, numerous academics have posited that Ubuntu embodies a robust spiritual foundation, serving as a comprehensive way of life that cultivates inner peace and a positive outlook on the world. It is an inherent orientation that possesses the remarkable capacity to inspire, encourage, and guide individuals towards feeling, perceiving, and acting in an inherently ethical manner towards their fellow human beings (Mawere & Mubaya, 2016; Stuit, 2016).

Individuals who actively practise or conscientiously maintain the core tenets of Ubuntu are generally expected to consistently demonstrate strong values, such as solidarity in times of need, unwavering mutual support among community members, a profound sense of togetherness that transcends individual differences, genuine brotherliness (or sisterliness) in their interactions, and a welcoming spirit of hospitality extended towards each other and indeed all members of the community in which they reside and interact (Stuit, 2016). The profound philosophy of Ubuntu has been consistently identified by a diverse range of authors and scholarly works as possessing a notably positive influence on the dynamics and outcomes of group work endeavours. Consequently, it is believed to positively impact the processes and fairness of group assessment methodologies (Barac et al., 2021; Beets, 2012; Chakabwata & Mukazi, 2022a, 2022b; Jili, 2022). Interestingly, despite the recognised positive associations, a thorough review of the existing academic literature has thus far yielded no documented instances or research findings that specifically indicate that Ubuntu has a negative impact on the crucial aspect of how students assess their peers within the context of collaborative group work. This apparent gap in the literature highlights a potential area for further investigation into the nuanced interplay between Ubuntu and peer assessments. It needs to be mentioned that although Ubuntu is a name claimed by African groups (Chakabwata & Mukazi, 2022a, 2022b), other cultural groups throughout the world also practise the essence of what Ubuntu stands for. These include working for the greater good of the group itself. This can be identified as the concept of shared intentionality. Despite the documented benefits of Ubuntu for group cohesion, its potential to inflate peer scores remains underexplored.

Shared intentionality represents a fundamental and uniquely human cognitive capacity that underpins our ability to engage in complex collaborative activities directed towards mutually understood goals and intentions. As prominently theorised by developmental psychologist Michael Tomasello and his colleagues, this capacity for shared intentionality is considered a defining characteristic that sets humans apart from even our closest primate relatives, such as chimpanzees (Wolf & Tomasello, 2025). While primates exhibit forms of cooperation and social interaction, shared intentionality goes beyond simple dyadic (two-person) collaboration with individual goals. It involves a sophisticated understanding of joint goals, shared commitments, and the ability to coordinate actions and intentions with others in a truly collective manner (O'Madagain & Tomasello, 2021).

The implications of shared intentionality for human social life are profound and far-reaching. It is considered a foundational building block for a wide range of complex social behaviours that are characteristic of human societies. For instance, group decision-making relies on the ability of individuals to understand shared goals, consider different perspectives, and collectively agree on a course of action (Chater et al., 2022). Therefore, shared intentionality is not merely a cognitive ability but a fundamental driver of human social complexity and cultural development (Tomasello & Carpenter, 2007).

Ubuntu emphasises communal values, interconnectedness, and collective responsibility, whereas Western individualism prioritises personal autonomy, self-expression, and individual achievements. These contrasting worldviews can create cultural tensions in assessment (Hofstede, 2011), where Ubuntu may not favour group-based evaluations, as they work towards shared success and will hesitate to point out the members of the group that did not participate. With Western models that often stress individual performance and competition, group members will be more willing to point out who did not contribute.

The philosophy of Ubuntu, which emphasises human interconnectedness and shared humanity, aligns closely with the objectives of SDG 4: Quality Education. Ubuntu's principle of inclusivity supports equitable access to education by valuing every learner's presence and potential, thereby addressing the goal of leaving no one behind (Adewale, 2023; Randle, 2024; Takyi-Amoako & Assié-Lumumba, 2018). Its focus on empathy and respect contributes to the creation of safe, non-violent learning environments in which students feel valued and supported (Adewale, 2023; Koen, 2021; Randle, 2024). Through community participation, Ubuntu encourages lifelong learning and active engagement beyond formal education, fostering a culture of collective growth and responsibility (Blose & Gumbo, 2024; Randle, 2024; Takyi-Amoako & Assié-Lumumba, 2018). Finally, Ubuntu's emphasis on holistic development ensures that education nurtures not only academic achievement but also emotional, social, and ethical growth, enhancing the overall quality of learning outcomes (Koen, 2021; Ogwo, 2018; Wuta, 2025). Together, these principles demonstrate how Ubuntu can serve as a culturally grounded framework for achieving inclusive, equitable, and transformative education, in line with SDG 4.

2.2 Methodology

In this research study, a mixed-methods approach (sequential explanatory design) was employed to provide a comprehensive understanding of why students did not truthfully score the participation of their peers. In the quantitative component of the study, score sheets were given to all group members to conduct peer group evaluations on the perceived participation of other group members. After observing the way all the students scored their peers' participation, it was necessary to delve deeper and determine why they scored as they did. To complement the quantitative data gathered through peer assessments, the qualitative component of the research involved semi-structured, one-on-one interviews with the students. The interviews were conducted during the same semester in which the students were registered for the subject in which the peer evaluations took place, ensuring contextual relevance and immediacy of reflection. The interviews, held in the researcher's office on campus, aimed to explore the reasoning behind students' scoring decisions, offering deeper insight into their perspectives and assessment criteria. Students were invited to participate voluntarily, and their responses were manually transcribed. Anonymity was maintained throughout the process by refraining from collecting any personally identifiable information, thereby ensuring ethical integrity and confidentiality of the participants. The target population for this study comprised all students enrolled in the Business Analysis course, a core component of an Information Technology qualification offered at a University of Technology located in South Africa. As all students participated in the study, concerns regarding representativeness or bias were minimised. As the entire population was included, no inclusion or exclusion criteria were applied. Additionally, the scores were not used for formal evaluation purposes but solely to support a deeper understanding of the phenomenon under investigation. Given the specific focus of the research on peer evaluations of participation within this particular course's group work activities, a purposive sampling method was deemed the most appropriate strategy. This approach ensured that the participants were directly relevant to the research question, as they were students who had engaged in group work and subsequent peer assessment processes under scrutiny (Campbell et al., 2020). Furthermore, the study achieved a whole population sampling strategy, meaning that all students enrolled in the Business Analysis course (48s-year students) were invited and actively participated in both the quantitative (submission of score sheets) and qualitative (participation in interviews) phases of the research. This comprehensive participation of the entire accessible population allowed for the achievement of data saturation, indicating that no new significant insights were obtained from further data collection.

In the context of assessing collaborative group assignments, a structured peer evaluation process was implemented to measure the individual contributions of each team member. As part of this evaluation, every student within the group was tasked with assessing their fellow group members by assigning a numerical participation score. To ensure consistency and clarity in the evaluation process, specific scoring criteria were provided to the students. According to these guidelines, a score of 100% was designated to indicate that a student had actively engaged in group work and had made a satisfactory and meaningful contribution to the overall completion of the assignment. Conversely, if a group member's contributions were perceived by their peers as being insufficient or below the expected standard, the students were instructed to assign a score of 1–50% for minimal contribution and 51–99% for moderate contribution. Furthermore, to account for instances of complete non-participation or a total lack of contribution from a group member, the scoring rubric stated that a score of 0% should be awarded. The study consisted of 12 distinct groups of students, with each group intentionally formed to consist of four individual members, facilitating a balanced and manageable structure for collaborative work and subsequent peer evaluation.

Individual, one-on-one interviews were conducted with the members of each student group as a crucial component of this study. The primary objective of these in-depth conversations was to gain an understanding of the reasoning behind a notable trend observed in the peer evaluations: the seemingly ubiquitous awarding by group members of 100% participation scores to all group members. This pattern was particularly intriguing given informal indications shared by students with the lecturer throughout the semester, suggesting that the level of participation among certain group members was significantly less than complete, with some reportedly not participating at all in the discussion. To systematically explore the underlying motivations and perceptions influencing this scoring behaviour, a manual thematic analysis was performed on qualitative data gathered from the interview transcripts. The following key themes emerged from the detailed comments and insights shared by the students during these informal, confidential one-on-one interviews:

1. Feelings of guilt.
2. Intimidation or bullying
3. Peer pressure
4. Shared intentionality
5. Ubuntu

2.3 Results

All 48 students scored their group members 100% for participation (Table 2.1).

The data presented in Table 2.1 demonstrate that the intended outcome of this peer evaluation exercise was not achieved. The primary purpose of implementing this assessment was to establish a mechanism for differentiating levels of student contribution within group work, thereby allowing for the appropriate recognition of active participants and, ultimately, the potential penalisation of those who do not contribute adequately to the group. However, the consistent awarding of 100% participation marks by all students to their group members, irrespective of their actual levels of engagement or contribution, effectively renders this exercise ineffective in fulfilling its intended purpose of peer evaluation. This uniform scoring pattern obscures any variations in individual effort and undermines the ability to accurately identify and address instances of non-participation, ultimately making the peer evaluation process, as implemented in this instance, largely obsolete as a tool for differentiating student contributions. During informal consultations with students during the semester, while they were still busy with the group assignment, it was mentioned that not all the group members participated. The limitations of the scoring results suggest that this method is insufficient for accurately identifying individual contributions to group projects. Consequently, alternative approaches may be necessary to assess student participation more effectively in the future.

Table 2.1 Marks distribution as given by students to peers

Number of students receiving a 100% participation score	Number of students receiving a Less than 100%, but more than 0% participation score	Number of students receiving a 0% participation score
48	0	0

Through semi-structured interviews with 21 students (all students who participated in scoring their peers did not avail themselves to participate in the interviews), the intention was to investigate and understand this discrepancy.

Students' experiences with peer assessment reveal a complex interplay of emotional, social, and cultural factors. Many expressed feelings of guilt when assigning low participation scores to peers who had not contributed meaningfully, fearing that they might jeopardise others' chances of passing. One student reflected, 'At the end of the day we all want to pass, and I wouldn't want to be the reason why others failed…', highlighting the tension between fairness and compassion. This was further complicated by intimidation and fear of conflict, with some students admitting that they gave full scores to avoid confrontation or disrupt group harmony. As one noted, 'Giving lower scores could lead to tension… it's a delicate balance between honesty and maintaining a positive working relationship'. Peer pressure also played a significant role, especially in groups formed among friends, where mutual agreements to award each other high marks were common. Students feared that honest scoring might damage relationships or create future friction, particularly in ongoing academic collaboration. Additionally, shared intentionality emerged, with some students sympathising with underperforming peers and believing that generous scoring might help them progress academically. Underpinning many of these decisions was the principle of Ubuntu, which emphasises collective well-being and solidarity. One student captured this sentiment, stating, 'We are for each other, not against each other… I would not want another student repeating a module or being left behind'. These insights illustrate how cultural values and interpersonal dynamics can significantly influence assessment practices, often challenging the objective evaluation of individual contributions.

Comments from all interviews were grouped into the above themes according to where they fit best. Figure 2.1 summarises the number of comments which fit each theme.

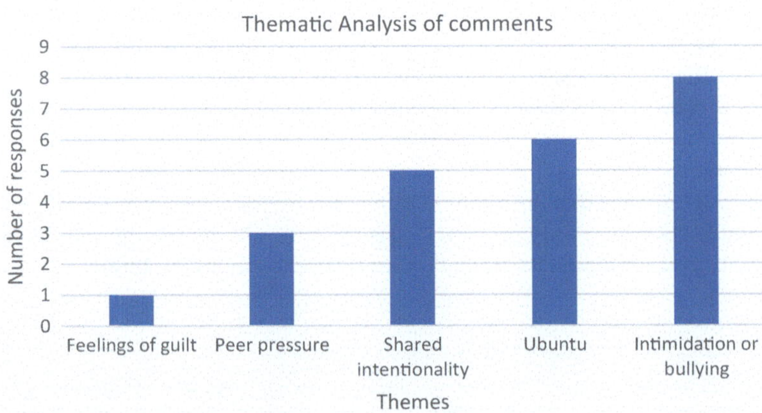

Fig. 2.1 Thematic analysis of comments

The thematic analysis of the interview data revealed several key factors influencing students' tendency to award all group members a perfect 100% participation score, regardless of actual contribution levels. The most prominent themes were intimidation or bullying, the influence of the philosophy of Ubuntu, and the impact of shared intentionality. To a lesser extent, peer pressure and feelings of guilt were also observed as contributing factors by a smaller subset of students.

Intimidation or Bullying: The finding that some students were hesitant to score group members lower than 100% because of fear of intimidation or bullying carries significant implications for the validity and fairness of peer assessment. This suggests a breakdown in the intended safe and objective evaluation environments. Students who feel threatened may inflate scores not to provide accurate feedback but to protect themselves from potential negative repercussions within the group. This undermines the formative and summative purposes of peer assessment, as it masks true levels of participation and does not provide meaningful feedback for improvement of the process. Educators implementing peer assessment must actively foster a climate of respect and safety by explicitly addressing and mitigating any potential bullying or intimidation within group dynamics. Strategies could include anonymous feedback mechanisms, clear guidelines for respectful communication, and proactive intervention by instructors when such issues are suspected.

The Philosophy of Ubuntu: The strong influence of Ubuntu, with its emphasis on interconnectedness, empathy, and communal well-being, presents a more complex dynamic. While Ubuntu is often lauded for its positive impact on group cohesion and support, our findings suggest that it can inadvertently hinder honest and critical peer evaluations. The deeply ingrained value of avoiding harm or disadvantaging others within the community may lead students to prioritise group harmony over accurate individual assessments. Assigning a lower score could be perceived as a violation of these Ubuntu principles, potentially causing interpersonal conflict or feelings of shame within the groups. This highlights a critical challenge in applying collectivist philosophies to individual assessment frameworks. Educators must find ways to harness the positive aspects of Ubuntu to foster collaboration while establishing clear expectations for honest and constructive feedback in peer assessment. This might involve explicit discussions about the purpose of peer assessment as a tool for collective growth, even if it requires acknowledging individual differences in contributions.

Theory of Shared Intentionality: The prevalence of shared intentionality as a contributing factor suggests that students' focus on the collective goal and shared responsibility for the group outcome may overshadow their individual assessment of participation. The strong desire to achieve the group's objective can lead students to view all members as equally invested in that goal, even if their actual contributions differ. This 'we are all in this together' mentality, while beneficial for group cohesion and motivation, can blur the lines of individual accountability in assessment. Students might be reluctant to differentiate contributions for fear of disrupting the perceived unity and shared ownership of a project. To address this, educators could implement strategies that promote individual accountability within group work, such as clearly defined roles and responsibilities, interim individual contribution reports, and assessment criteria that specifically target individual effort and engagement alongside the final group product.

Peer Pressure and Feelings of Guilt: While observed by fewer students, the presence of peer pressure and feelings of guilt further underscored the social complexities inherent in the peer assessment. Students might feel pressured by their peers to award high scores, particularly in close-knit groups, where there is a strong desire for social acceptance. Similarly, students who may have contributed less themselves might feel guilty about scoring others lower, fearing reciprocal negative evaluations or social disapproval. These factors highlight the importance of creating a peer assessment process that minimises social pressures and encourages objective evaluation based on clearly defined criteria rather than interpersonal dynamics. Anonymity, clear instructions emphasising the importance of honest feedback for learning, and instructor guidance on providing constructive criticism can help to mitigate these influences.

The results of this study provide valuable insights into the social and cultural factors that affect peer assessment practices. These findings are directly relevant to achieving Sustainable Development Goal 4 (SDG 4), which seeks to ensure inclusive and equitable quality education and promote lifelong learning opportunities for everyone. This study shows that although peer assessment is acknowledged as an effective educational tool for promoting collaboration, accountability, and critical thinking (Butarbutar et al., 2025; Li et al., 2020), its efficacy can be undermined by cultural and interpersonal elements such as peer pressure, intimidation, and the philosophy of Ubuntu(Barac et al., 2021; Burke Moneypenny et al., 2018; Raes et al., 2015). As a result of these factors, students might give inflated participation scores, which compromises the reliability of peer assessment as a tool for evaluating individual contributions and fostering fairness in the assessment process.

For example, the issue of intimidation or bullying underscores a major obstacle to establishing safe and non-violent educational settings, one that is explicitly aimed at by SDG 4. The fact that students are hesitant to give honest scores to their peers because they fear conflict or retaliation indicates that the learning environment might not provide sufficient support or psychological safety. This is consistent with the results of earlier investigations (Bursztyn & Jensen, 2015; Hill & Zhou, 2023; Zhou et al., 2020), which highlighted the crucial roles of trust and transparency in peer assessment procedures. To support the views of SDG 4, institutions need to establish protective measures, such as anonymous feedback mechanisms and explicit behavioural protocols. These measures will shield students from coercion and guarantee that evaluations are based on real involvement and exertion of effort.

Although the impact of Ubuntu is a source of cultural enrichment, it also brings with it a subtle challenge. Ubuntu's focus on empathy, solidarity, and collective success may clash with the personal responsibility needed in assessment situations (Davison, 2023). Similar tensions have been noted in collectivist cultures, where group harmony often prevails over individual critiques (Jasim, 2022; Triandis, 2001).

Although Ubuntu promotes inclusivity and mutual support, which are fundamental principles of SDG 4, it may inadvertently deter honest evaluation, which could impact the quality of learning outcomes. Thus, teachers must find a compromise between combining the positive aspects of Ubuntu with cooperative learning and emphasising the value of constructive, individualised feedback for personal and academic growth.

The notion of shared intentionality adds complexity to the assessment landscape. The inclination of students to regard all group members as equally dedicated to the shared objective, irrespective of their actual contributions, demonstrates a robust sense of group ownership. This fosters motivation and unity but may conceal differences in effort and contribution. This discovery aligns with studies on social loafing and groupthink (Dahiam & Ghaleb, 2024; Tosuntaş, 2020), which warn against undermining individual responsibility in group contexts. In line with SDG 4's focus on quality education, teachers are encouraged to explore hybrid assessment models that integrate group evaluations and individual performance metrics to ensure a precise representation of each student's learning journey.

Ultimately, although they are not as prominent, the influences of guilt and peer pressure highlight the emotional intricacies of peer assessment. Such elements can skew students' assessment of situations and put the integrity of the evaluation process at risk. Previous research by Barac et al. (2021) similarly observed that emotional and relational dynamics can impact assessment behaviour, especially within close-knit groups. Confronting these challenges necessitates a conscious endeavour to develop students' assessment literacy, highlighting the importance of sincere feedback and the contribution of peer assessment to improvement rather than penalisation of learning.

2.4 Conclusion

The findings of this thematic analysis reveal a multifaceted set of reasons for the inflation of peer assessment scores for group participation. Addressing this issue requires a nuanced approach that considers not only the practical design of the assessment process but also the underlying social dynamics, cultural values, and psychological factors at play within student groups. Educators must actively cultivate a learning environment that balances collaboration and individual accountability, promotes psychological safety for honest feedback, and clarifies the purpose and value of accurate peer assessment for both individual and collective growth.

Although it might be tempting to assign the failed outcome of the scoring of group members' participation to Ubuntu, it cannot be done blindly.

Peer group evaluation has great potential, but it is not without its difficulties. It can be challenging to find a balance between offering constructive criticism and preserving positive group dynamics. However, these difficulties also offer opportunities for development and advancement (Adachi et al., 2018), and professionals and educators should consider creative frameworks and methods that deal with such issues.

Further research might be necessary if this experiment is repeated to determine whether Ubuntu influences the honest scoring of group members. It might be insightful to determine whether the same would have happened in a purely Western setting.

In conclusion, a careful examination of the relationship between individual contributions and group dynamics is required to determine the elements that contribute to a good peer group review.

We can create a more sophisticated, just, and empowering collaborative experience by resolving the complexities of peer-group evaluation. For further research, it might be interesting to conduct a similar study to determine whether this tendency to not be willing to evaluate fellow students' participation justly is also present in purely Western culture groups or if this behaviour is solely motivated by the African Ubuntu ideology. Future research should focus on other cultural groups. This could provide information that will help guide best practices and add to the continuing discussion on developing efficient peer assessment techniques.

This study adds to the growing body of literature on assessment methodologies by elucidating the complex relationships between Ubuntu philosophy and peer group evaluation. It also illuminates the possibility of incorporating cultural values into professional and educational practice. Finally, knowing how Ubuntu affects peer assessment efficacy may provide insights into creating inclusive, cooperative, and compassionate learning and work cultures.

References

Adachi, C., Tai, J.H.-M., & Dawson, P. (2018). Academics' perceptions of the benefits and challenges of self and peer assessment in higher education. *Assessment & Evaluation in Higher Education, 43*(2), 294–306. https://doi.org/10.1080/02602938.2017.1339775

Adewale, S. (2023). Promoting sustainable development goal 5 through ubuntu-based South African basic education curriculum. *Studies in Learning and Teaching, 4*(3), 432–443. https://doi.org/10.46627/silet.v4i3.231

Akabor, S., & Phasha, N. (2025). Where is Ubuntu in competitive South African schools? An inclusive education perspective. *International Journal of Inclusive Education, 29*(3), 361–377. https://doi.org/10.1080/13603116.2022.2127491

Ashenafi, M. M. (2017). Peer-assessment in higher education—twenty-first century practices, challenges and the way forward. *Assessment & Evaluation in Higher Education, 42*(2), 226–251. https://doi.org/10.1080/02602938.2015.1100711

Babo, R., Rocha, J., Fitas, R., Suhonen, J., & Tukiainen, M. (2021). Self and peer e-assessment: A study on software usability. *International Journal of Information and Communication Technology Education (IJICTE), 17*(3), 68–85. https://doi.org/10.4018/IJICTE.20210701.oa5

Barac, K., Kirstein, M., & Kunz, R. (2021). Using peer review to develop professional competencies: An Ubuntu perspective. *Accounting Education*. https://www.tandfonline.com/doi/abs/10.1080/09639284.2021.1942089

Beets, P. A. D. (2012). Strengthening morality and ethics in educational assessment through Ubuntu in South Africa. *Educational Philosophy and Theory, 44*(sup2), 68–83. https://doi.org/10.1111/j.1469-5812.2011.00796.x

Blose, P., & Gumbo, M. T. (2024). Developing a framework for integrating IKS in technology education with sustainable development principles. *African Journal of Research in Mathematics, Science and Technology Education, 28*(3), 330–342. https://doi.org/10.1080/18117295.2024.2356416

Burke Moneypenny, D., Evans, M., & Kraha, A. (2018). Student perceptions of and attitudes toward peer review. *American Journal of Distance Education, 32*(4), 236–247. https://doi.org/10.1080/08923647.2018.1509425

Bursztyn, L., & Jensen, R. (2015). How does peer pressure affect educational investments? *The Quarterly Journal of Economics, 130*(3), 1329–1367. https://doi.org/10.1093/qje/qjv021

Butarbutar, R., Sauhenda, A. F., & Marnina. (2025). A systematic review of online peer assessment tools for learning evaluation. *Innovare Journal of Education*, 1–7. https://doi.org/10.22159/ijoe.2025v13i2.53640

Campbell, S., Greenwood, M., Prior, S., Shearer, T., Walkem, K., Young, S., Bywaters, D., & Walker, K. (2020). Purposive sampling: Complex or simple? Research case examples. *Journal of Research in Nursing, 25*(8), 652–661. https://doi.org/10.1177/1744987120927206

Casanova, M. (2022). Group work as a learning strategy in higher education. *Global Journal of Human-Social Science*, 9–17. https://doi.org/10.34257/GJHSSGVOL22IS5PG9

Chakabwata, W., & Mukazi, F. M. (2022a). Ubuntu philosophy and online assessment in higher education institutions. In *Handbook of research on transformative and innovative pedagogies in education* (pp. 257–275). IGI Global Scientific Publishing. https://doi.org/10.4018/978-1-7998-9561-9.ch014

Chakabwata, W., & Mukazi, F. M. (2022b). Ubuntu philosophy and online assessment in higher education institutions. In J. Keengwe (Ed.), *Advances in educational technologies and instructional design* (pp. 257–275). IGI Global. https://doi.org/10.4018/978-1-7998-9561-9.ch014

Chater, N., Zeitoun, H., & Melkonyan, T. (2022). The paradox of social interaction: Shared intentionality, we-reasoning, and virtual bargaining. *Psychological Review, 129*(3), 415–437. https://doi.org/10.1037/rev0000343

Cheng, W., & Warren, M. (2000). Making a difference: Using peers to assess individual students' contributions to a group project. *Teaching in Higher Education, 5*(2), 243–255. https://doi.org/10.1080/135625100114885

Dahiam, B., & Ghaleb, S. (2024). Social loafing: Understanding, mitigating, and enhancing group performance. *International Journal of Scientific Multidisciplinary Research, 2*(9), 1321–1328. https://doi.org/10.55927/ijsmr.v2i9.10975

Davison, Z. (2023). Freedom for responsibility: The essence of Ubuntu/Unhu philosophy. *Philosophy International Journal, 6*(3), 1–9. https://doi.org/10.23880/phij-16000305

De Castro, V. B., Sridharan, B., Watty, K., & Safari, M. (2021). The impact of learner engagement on performance outcomes: A longitudinal study in accounting education. *Accounting & Finance, 61*(S1), 1639–1664. https://doi.org/10.1111/acfi.12640

De Meulemeester, A., Peleman, R., Pauwels, N. S., & Buysse, H. (2022). Peer assessment, self-assessment and teacher scoring within an information literacy course. In S. Kurbanoğlu, S. Špiranec, Y. Ünal, J. Boustany & D. Kos (Eds.), *Information literacy in a post-truth era* (Vol. 1533, pp. 490–501). Springer International Publishing. https://doi.org/10.1007/978-3-030-99885-1_41

Driskell, J. E., Salas, E., & Driskell, T. (2018). Foundations of teamwork and collaboration. *American Psychologist, 73*(4), 334–348. https://doi.org/10.1037/amp0000241

Freeman, M. (1995). Peer assessment by groups of group work. *Assessment & Evaluation in Higher Education, 20*(3), 289–300. https://doi.org/10.1080/0260293950200305

Gaynor, J. W. (2020). Peer review in the classroom: Student perceptions, peer feedback quality and the role of assessment. *Assessment & Evaluation in Higher Education, 45*(5), 758–775. https://doi.org/10.1080/02602938.2019.1697424

Hall, D., & Buzwell, S. (2013). The problem of free-riding in group projects: Looking beyond social loafing as reason for non-contribution. *Active Learning in Higher Education, 14*(1), 37–49. https://doi.org/10.1177/1469787412467123

Hernández González, C. A., & Blackford, B. J. (2022). Engagement as antecedent of academic achievement and the moderating impact of work-family-school inter-role conflict for online graduate students. *The International Journal of Management Education, 20*(3), Article 100676. https://doi.org/10.1016/j.ijme.2022.100676

Hill, A. J., & Zhou, W. (2023). Peer discrimination in the classroom and academic achievement. *Journal of Human Resources, 58*(4), 1178–1206. https://doi.org/10.3368/jhr.59.2.0919-10460R3

Hofstede, G. (2011). Dimensionalizing cultures: The Hofstede model in context. *Online Readings in Psychology and Culture, 2*(1). https://doi.org/10.9707/2307-0919.1014

Hu, Y., Nath, N., Zhu, Y., & Laswad, F. (2024). Accounting students' online engagement, choice of course delivery format and their effects on academic performance. *Accounting Education, 33*(5), 649–684. https://doi.org/10.1080/09639284.2023.2254298

Jasim, A. M. (2022). Collectivism VS individualism in H. G. Wells's Kipps (1905) and Edith Wharton's the age of innocence (1920). *Journal of STEPS for Humanities and Social Sciences, 1*(3). https://doi.org/10.55384/2790-4237.1100

Jili, N. N. (2022). Responsive assessment tools in an inclusive education setting. In M. O. Maguvhe & M. M. Masuku (Eds.), *Advances in educational technologies and instructional design* (pp. 195–206). IGI Global. https://doi.org/10.4018/978-1-6684-4436-8.ch015

Kench, P. L., Field, N., Agudera, M., & Gill, M. (2009). Peer assessment of individual contributions to a group project: Student perceptions. *Radiography, 15*(2), 158–165. https://doi.org/10.1016/j.radi.2008.04.004

Kennedy, G. J. (2005). Peer-assessment in group projects: Is it worth it? *7th Australasian Conference on Computing Education, 42*, 59–65.

Koen, M. (2021). Sustainable future for early childhood: Applying the African Ubuntu philosophy to contribute to the holistic development of young children. In W. Leal Filho, R. Pretorius & L. O. De Sousa (Eds.), *Sustainable development in Africa* (pp. 131–146). Springer International Publishing. https://doi.org/10.1007/978-3-030-74693-3_8

Lejk, M., & Wyvill, M. (2002). Peer assessment of contributions to a group project: Student attitudes to holistic and category-based approaches. *Assessment & Evaluation in Higher Education, 27*(6), 569–577. https://doi.org/10.1080/0260293022000020327

Li, H., Xiong, Y., Hunter, C. V., Guo, X., & Tywoniw, R. (2020). Does peer assessment promote student learning? A meta-analysis. *Assessment & Evaluation in Higher Education, 45*(2), 193–211. https://doi.org/10.1080/02602938.2019.1620679

Linn, Z., Tashiro, Y., Morio, K., & Hori, H. (2022). Peer evaluations of group work in different years of medical school and academic achievement: How are they related? *BMC Medical Education, 22*(1), 102. https://doi.org/10.1186/s12909-022-03165-5

Mahat, A., Md Supie, H. S., Ahmad, N. I. S., Zainal Abidin, N. A., Jama, S. R., & Michael, Z. (2024). Exploring the relationship between all stages in group work. *International Journal of Academic Research in Business and Social Sciences, 14*(7), 392–405. https://doi.org/10.6007/IJARBSS/v14-i7/21942

Mawere, M., & Mubaya, T. R. (2016). *African philosophy and thought systems: A search for a culture and philosophy of belonging*. Langaa RPCIG.

Ogwo, B. A. (2018). Re-visioning technical vocational education and training (TVET) for the youth in Sub-Saharan Africa (SSA) and the sustainable development goals (SDGs): Prospects and promises within the framework of the Ubuntu paradigm. In E. J. Takyi-Amoako & N. T. Assié-Lumumba (Eds.), *Re-visioning education in Africa* (pp. 155–173). Springer International Publishing. https://doi.org/10.1007/978-3-319-70043-4_9

O'Madagain, C., & Tomasello, M. (2021). Shared intentionality, reason-giving and the evolution of human culture. *Philosophical Transactions of the Royal Society b: Biological Sciences, 377*(1843), 20200320. https://doi.org/10.1098/rstb.2020.0320

Raes, A., Vanderhoven, E., & Schellens, T. (2015). Increasing anonymity in peer assessment by using classroom response technology within face-to-face higher education. *Studies in Higher Education, 40*(1), 178–193. https://doi.org/10.1080/03075079.2013.823930

Rajabzadeh, A. R., Long, J., Saini, G., & Zeadin, M. (2022). Engineering student experiences of group work. *Education Sciences, 12*(5), 288. https://doi.org/10.3390/educsci12050288

Randle, O. (2024). Providing Ubuntuism through participatory culture framework: Sustainable development goals 4 and 5. *Studies in Learning and Teaching, 5*(1), 140–151. https://doi.org/10.46627/silet.v5i1.356

Sefotho, M. M., & Letseka, M. (2024). Botho/Ubuntu paradigm as cognitive justice in psychology. *School Psychology International, 45*(3), 233–253. https://doi.org/10.1177/01430343241244748

Shah-Fairbank, S., Woo, J., Lamb, K., & Cho, J. (2021). Students' teamwork assessment based on reflection, peer evaluations and psychological safety. In *2021 ASEE Virtual Annual Conference Content Access Proceedings*, 37774. https://doi.org/10.18260/1-2--37774

Steenkamp, G., & Brink, S. M. (2024). Students' experiences of peer learning in an accounting research module: Discussion forums, peer review and group work. *The International Journal of Management Education, 22*(3), Article 101057. https://doi.org/10.1016/j.ijme.2024.101057

Stenberg, M., Mangrio, E., Bengtsson, M., & Carlson, E. (2021). Formative peer assessment in higher healthcare education programmes: A scoping review. *British Medical Journal Open, 11*(2), Article e045345. https://doi.org/10.1136/bmjopen-2020-045345

Stuit, H. (2016). *Ubuntu strategies: Constructing spaces of belonging in contemporary South African culture*. Springer.

Takyi-Amoako, E. J., & Assié-Lumumba, N. T. (2018). Conclusion: Towards an Ubuntu-inspired continental partnership on education for sustainable development in Africa—African Union Commission Agenda 2063 education strategy. In E. J. Takyi-Amoako & N. T. Assié-Lumumba (Eds.), *Re-visioning education in Africa* (pp. 229–246). Springer International Publishing. https://doi.org/10.1007/978-3-319-70043-4_13

Tomasello, M., & Carpenter, M. (2007). Shared intentionality. *Developmental Science, 10*(1), 121–125. https://doi.org/10.1111/j.1467-7687.2007.00573.x

Tosuntaş, Ş. B. (2020). Diffusion of responsibility in group work: Social loafing. *Journal of Pedagogical Research, 4*(3), 344–358. https://doi.org/10.33902/JPR.2020465073

Triandis, H. C. (2001). Individualism-collectivism and personality. *Journal of Personality, 69*(6), 907–924. https://doi.org/10.1111/1467-6494.696169

Tumpa, R. J., Skaik, S., Ham, M., & Chaudhry, G. (2022). A holistic overview of studies to improve group-based assessments in higher education: A systematic literature review. *Sustainability, 14*(15), 9638. https://doi.org/10.3390/su14159638

Van Gennip, N. A. E., Segers, M. S. R., & Tillema, H. H. (2009). Peer assessment for learning from a social perspective: The influence of interpersonal variables and structural features. *Educational Research Review, 4*(1), 41–54. https://doi.org/10.1016/j.edurev.2008.11.002

Watson, W. E., BarNir, A., & Pavur, R. (2010). Elements influencing peer evaluation: An examination of individual characteristics, academic performance, and collaborative processes. *Journal of Applied Social Psychology, 40*(12), 2995–3019. https://doi.org/10.1111/j.1559-1816.2010.00690.x

Wolf, W., & Tomasello, M. (2025). A shared intentionality account of uniquely human social bonding. *Perspectives on Psychological Science, 20*(2), 264–275. https://doi.org/10.1177/17456916231201795

Wuta, R. (2025). The sustainability of Unhu/Ubuntu as a philosophy of education in a postcolonial and globalising Zimbabwe: A hermeneutical discussion. *Journal of Indigenous Social Development, 13*(2), 97–120. https://doi.org/10.55016/ojs/jisd.v13i2.79623

Zhou, J., Zheng, Y., & Tai, J.H.-M. (2020). Grudges and gratitude: The social-affective impacts of peer assessment. *Assessment & Evaluation in Higher Education, 45*(3), 345–358. https://doi.org/10.1080/02602938.2019.1643449

Chapter 3
Reflective Teaching Through Character Strengths: A Self-study of Reciprocal Educator–Student Growth in the Teaching Advancement of Universities (TAU) Programme

Fathima Essop Mahomed, Anthony K. Isabirye, and Carmel Proctor

Abstract This self-study examines reciprocal educator-student growth arising from a strengths-based pedagogy grounded in positive psychology as part of an institutional programme within the national Teaching Advancement at University (TAU) Fellowship Programme. Positioning the practitioner-researcher as a participant, the study traced iterative enactment reflection cycles with a cohort of 103 first-year Human Resource Management students, combining structured Gibbs-aligned reflection with Mezirowian criteria for transformative learning. Multiple qualitative artefacts (researcher journals, lesson plans, audio-visual stimulated recall, student reflective logs and essays, field notes, and programme documents) were triangulated and interrogated with critical friends to broaden the interpretive vantage points and enhance rigour. An analytic protocol mapped Gibbs' reflective micro-stages (description, feelings, evaluation, analysis, conclusion, action) onto Mezirow's indicators of transformative learning (disorienting dilemmas, critical reflection, communicative testing, and enacted transformation) to distinguish between instrumental adjustments and perspective change. The findings coalesced into six interrelated themes. Three of these themes are related to educator development: (1) teacher identity and reflective practice; (2) pedagogical innovation, instructional redesign, and (3) leadership, agency, and social justice advocacy. The other three focus on student development: (1) strengths-based self-awareness and agency; (2) collaboration, interpersonal growth, and academic engagement; and (3) resilience, and professional readiness. Evidence of mutual reciprocity emerged when teacher interventions generated disorienting

F. E. Mahomed (✉) · A. K. Isabirye
Faculty of Management Sciences, Vaal University of Technology, Vanderbijlpark, South Africa
e-mail: fathima@vut.ac.za

A. K. Isabirye
e-mail: isabiryeaki@gmail.com

C. Proctor
Positive Psychology Research Centre, St. Saviour, Guernsey
e-mail: carmel@pprc.gg

K. Dube and L. M. Tjabane (eds.), *Advancing Sustainable Development Goals in Higher Education*, https://doi.org/10.1007/978-3-032-27646-9_3

experiences for students, prompting critical reflection, which, in turn, prompted the educator to re-examine their practice and sustain change for the benefit of all involved in the learning experience. These implications underscore the value of integrated, reflective frameworks, embedded character-strengths activities, and institutional support for reciprocal, equity-oriented professional development. Recommendations for programme design and future multi-site research are offered to test the transferability of the Reciprocal Growth Model.

Keywords Reflective teaching · Character strengths · Strengths-based pedagogy · TAU higher education · Strengths gym

3.1 Introduction

The Teaching Advancement at University (TAU) Programme, established by the Higher Education Learning and Teaching Association of Southern Africa (HELTASA), represents a coordinated national effort to advance educational leadership, foster innovative pedagogies and promote inclusive teaching practices across South Africa. Responsive to shifting student demographics and institutional imperatives, TAU prioritises the professional growth of its academic staff as a catalyst for systemic transformation in higher education. As a TAU participant, the researcher in this case served as a change agent by promoting strengths-based and transformative methods in higher education with positive psychology.

Despite the global popularity of positive psychology and strengths-based interventions, research on the experiences and professional growth of educators who utilise strengths-based pedagogy remains scarce. Primarily focusing on student outcomes, the literature overlooks how these approaches might stimulate changes in educator identity, reflective practice, and institutional and classroom transformation, particularly in the distinct socio-educational context of South African universities.

This study aims to address this critical gap by conducting a reflective self-study within the TAU Programme, focusing on the implementation of Level 3 of the *Strengths Gym* programme. *Strengths Gym* (Proctor & Fox Eades, 2019) is a positive psychology, strength-based intervention aligned with the Values in Action (VIA) Institute, a non-profit organisation dedicated to the science and practice of character strengths (Peterson & Seligman, 2004). Character strengths comprise 24 universal positive qualities that represent the best of human nature. These strengths are categorised into six virtues: Wisdom and Knowledge, Courage, Humanity, Justice, Temperance, and Transcendence. Individuals can discover their character strengths through the VIA Survey, which helps foster personal growth, improve well-being, and build resilience by aligning one's actions with core values. *Strengths Gym's* structured, dialogic, and experiential activities are designed to help individuals identify, apply, and reflectively utilise their character strengths.

Underpinned by Mezirow's Transformative Learning Theory (TLT) and bolstered by Gibbs' reflective cycle, this investigation explores how a strengths-based intervention within TAU can promote mutual growth, also referred to as reciprocal growth, for teachers and students. Reciprocal Growth is understood as a dynamic principle within complex systems, where synergistic interactions between components drive collective advancement and sustained well-being. The development of teaching methods, changes in professional identities, and promotion of inclusivity were all considered. The central research question that frames this inquiry is, "*How can a strengths-based intervention within the TAU Programme foster simultaneous educator and student development in a South African university context?*"

3.2 Literature Review

3.2.1 Strengths-Based Pedagogy in Higher Education

The concept of "strength-based pedagogy" originates from a range of educational and psychological traditions and does not have a single originator (Fung, 2017). Although the term "strengths perspective" was first used in social work by Weick et al. (1989) to highlight human autonomy and potential, Clifton and Nelson's (1992) work significantly advanced its application in education, and Peterson and Seligman (2004) formalised it using positive psychology principles. Using frameworks such as the VIA Classification to increase academic engagement, resilience, and subjective well-being, modern strengths-based programmes include these theoretical underpinnings to promote "optimal functioning" in learners. This method focuses on identifying what students can do well, building on their real strengths rather than the problems they face. Learning spaces shaped this way help students feel connected, make decisions for themselves, and grow their confidence in their abilities. Instruction is shifting toward matching students' interests and skills (Modry-Mandell & Nguyen, 2022). When applied in daily teaching, the approach helps learners who face various challenges, such as those with disabilities. Support often comes through lesson plans rooted in personal interests, aiming to improve both school achievement and overall well-being. Advocates of this approach argue that by focusing on strengths, educators can enhance engagement, build confidence and resilience, and co-construct more inclusive and equitable classroom partnerships (Fung, 2017; Modry-Mandell & Nguyen, 2022).

This perspective is particularly valuable in South African higher education, where first-generation and historically disadvantaged students would benefit from transformative pedagogies that foreground students' existing skills and contextual resources, thereby contributing to the broader social justice agenda of educational equity and participation parity rather than the deficit approach that focuses on limitations. (Leibowitz, 2017). Strengths-based pedagogy has been shown to promote engagement, self-efficacy, resilience, and well-being by highlighting what students can do and providing structured opportunities to recognise, apply, and reflect on their character strengths, such as through the *Strengths Gym* intervention (Lopez & Louis, 2009; Linkins et al., 2015; Soria & Stebleton, 2012). Along with supporting changes in educator practice, these interventions help instructors transition from being content transmitters to facilitators who focus on the overall development of their students and align classroom practices with the ideals of social justice and inclusion. Incorporating strengths-based activities into the regular curriculum ultimately helps students learn while also allowing teachers to improve professionally, as both groups engage in cycles of self-reflection, self-application and self-identification. Furthermore, the empowering feature of this transformative pedagogy contributes to the broader goal of achieving quality education, as outlined in Sustainable Development Goal 4 (SDG 4: ensure inclusive and equitable quality education and promote lifelong learning opportunities for all). SDG 4 emphasises the crucial role of access to quality education, ensuring access to education for all, regardless of age, gender, ethnicity or disability. SDG 4 focuses not only on enrolment rates but also on enhancing the quality of education, emphasising that increased access requires additional investment in educational resources (A4ID, 2022).

Strength-based pedagogy is enacted through programmes such as *Strengths Gym*, which help students identify and apply their strengths (Proctor & Fox Eades, 2019). When incorporated into development programmes such as TAU, these interventions not only enhance student learning but also offer faculty opportunities for reflection and growth (de Kaat, 2020). Additionally, integrating strength-based pedagogy with social justice frameworks promotes equitable well-being. It encourages educators and students to critically examine their assumptions and adjust their personal and professional identities in the process. A strength-based approach fosters an ideal environment for transformative learning. Mezirow's Transformative Learning Theory provides a thorough theoretical framework for this process.

3.2.2 Transformative Learning Theory

Transformative Learning Theory (TLT), initially formulated by Mezirow (1991), views adult learning as a process of scrutinising and modifying fundamental presumptions, resulting in significant changes in viewpoint, attitudes, and behaviour. Experiences with unsettling situations, critical self-reflection, rational discourse, and adopting new ways of thinking and doing are the four interconnected processes at the heart of TLT. Because it places more emphasis on changing perspectives than on making technical adjustments, this framework offers a potent lens for professional development in higher education (Kitchenham, 2008). In response to the complex educational landscape of South Africa, TLT was operationalised in the TAU programme through action learning, collaborative enquiry, and structured reflection. This allowed for inclusive pedagogical adjustments and reframing of teacher identity. By securing cycles of transformation in discussion and introspection practice, TLT empowered educators to move beyond content delivery towards facilitation that addresses both the cognitive and affective dimensions of learning, with direct relevance for equity and social justice in post-apartheid contexts.

Building on this theoretical foundation, the practice of reflective teaching, particularly as enacted through systematic self-study, offers a concrete methodology for educators to engage with and enact transformative learning processes in their professional contexts.

3.2.3 Reflective Teaching and Self-study

Reflective teaching is a methodical and critical analysis of one's own teaching methods, combining academic understanding with practical insights to improve practice and enhance the positive student experience (Gibbs, 1988; Schön, 1983). Gibbs' Reflective Cycle—moving from description to feelings, evaluation, analysis, conclusion, and action plan—offers an organised structure for this work, prompting teachers to attend to both what happened and their emotional and intellectual responses to it. Embedded in reflective teaching, and particularly valuable for classroom-based enquiry, are the core elements of self-study: systematic documentation (e.g. reflective journals, lesson artefacts, recordings), multiple sources of data and triangulation (student feedback, peer observation, assessment outcomes), explicit theoretical framing that links practice to literature, reflexive interrogation of assumptions and positionality, transparent and ethical reporting of methods and findings, and iterative cycles of enactment and revision that both improve practice and generate knowledge (Pinnegar & Hamilton, 2009; Samaras, 2011). Together, these elements move reflection beyond impressionistic commentary toward rigorous, shareable inquiry that supports both individual development and collective learning.

Self-study is a methodical, evidence-based investigation into one's own teaching, with the dual goals of enhancing practice and contributing to academic discourse (Pinnegar & Hamilton, 2009). It emphasises transparency, reflexivity, and the integration of personal experience with scholarly frameworks (Samaras, 2011). Gibbs' Reflective Cycle provides a practical mechanism for operationalising the abstract principles of TLT, as its staged prompts encourage the critical reflection and testing of assumptions that lie at the heart of transformation (Cranton, 2023). For example, the cycle's analysis and conclusion phases invite teachers to surface and scrutinise their underlying beliefs about teaching and learning. In contrast, the action plan phase supports experimentation with alternative practices—steps that mirror transformative processes such as experiencing a disorienting dilemma, engaging in critical reflection, and trying out new perspectives. This article defines self-study as a methodical enquiry into my teaching using TLT and Gibbs' Reflective Cycle. It includes various data sources such as reflective journals, lesson artefacts, and student reflections. Through iterative cycles of reflection and engagement with critical friends, this study assesses how strengths-based pedagogy during TAU influenced my professional identity and fostered students' self-awareness, collaboration, and academic engagement.

3.2.4 *Social Justice in Pedagogy*

Social justice in education refers to the ongoing pursuit of equity and inclusion and the dismantling of systemic barriers that limit access, participation, and outcomes for historically marginalised students (Bali, 2020; Leibowitz, 2017). In the South African context, the legacy of apartheid and persistent structural inequalities have necessitated policies and pedagogies that foreground non-discrimination, redress, and redistribution of educational opportunities (Department of Education, 1997, 2001; Pendlebury & Enslin, 2004). Social justice frameworks in higher education demand not only the equitable distribution of resources but also the fostering of participatory, affirming learning environments that recognise and build upon the diverse strengths, identities, and lived realities of all learners. When paired with strengths-based pedagogy and transformative reflection, social justice commitments become actionable: educators are called to challenge deficit narratives, design inclusive curricula, and ensure that university spaces enable all students to flourish as active participants in shaping a democratic society. Numerous studies have identified various elements linked to students' disengagement experiences in higher education. Health, social support, university culture, curriculum, and identity (e.g. race and gender) are all important factors (Thomas, 2010; Wilcox et al., 2005; Yorke & Thomas, 2003). Additionally, students who become disengaged or marginalised because of unresponsive university practices are a common problem for university educators (Ryan, 2011). How educators engage their diverse students and acknowledge and build upon the resources they bring to their studies is critical to success in terms of social justice and student retention. Within the TAU Programme, exposure to socially engaged pedagogy challenged my assumptions about the power dynamics between lecturers and students and the broader university culture. A social justice perspective seeks not only individual growth but also equity, inclusion, and empowerment in educational environments. Strengths-based pedagogy inherently affirms diverse identities and backgrounds, recognising that all students bring valued traits to the learning space.

A comprehensive framework for change that promotes both academic achievement and individual well-being can be obtained by combining social justice and strengths-based approaches.

The synergistic model guiding this study is illustrated in Fig. 3.1. By recognising and utilising individual resources for development, strengths-based treatments lay the groundwork for bringing students' and educators' varied potential to light. The theory of transformative learning justifies and means profound transformation, encouraging critical introspection and shifts in perspective as people confront preconceived notions and consider novel options. A systematic and methodical approach to this transformation is provided by Gibbs' Reflective Cycle, which leads participants step-by-step through the process of characterising experiences, assessing results, and creating plans for positive action. Meanwhile, social justice grounds the process in an ethical imperative, ensuring that transformation is oriented toward equity, inclusion, and the empowerment of all participants, explicitly addressing the need for just and accessible educational outcomes. Together, these frameworks create a mutually

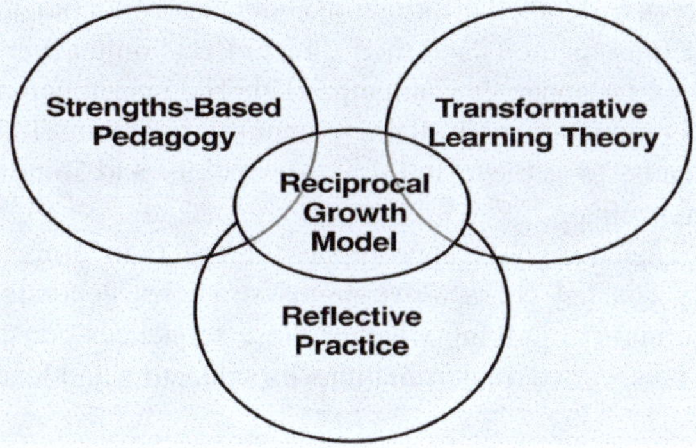

Fig. 3.1 Synergistic model. *Source* authors' own, with assistance of perplexity AI

reinforcing system wherein strengths-based activities initiate reflective, transformational processes, structured reflection refines and extends these insights, and resultant growth cycles back to inform a classroom culture that prioritises equity and inclusion, thus fully operationalising social justice as a core principle of daily academic practice.

3.3 Methodology

This research investigated reciprocal educator–student growth arising from a strength-based intervention informing reflective teaching within a national Teaching Advancement programme. A self-study design locates the practitioner–researcher as both participant and primary data source, adopting an interpretivist epistemology: knowledge is co-constructed through situated reflection and dialogic exchange rather than being discovered as an objective fact (Tolich, 2010; Lassonde & Galman, 2019). Given the focus on meaning change and professional practice, the study intentionally combines two complementary conceptual resources, Mezirow's TLT and a staged reflective procedure derived from Gibbs' reflective cycle, so that analytic claims can speak to both the micro-processes of reflection and more profound shifts in participants' meaning perspectives (Aubrey et al., 2015; Gibbs model operationalised via contemporary applications; Hashim et al., 2023; Ocampo et al., 2025). This section describes the design, context, data sources, and the integrated analytic strategy that maps Gibbs' stages onto Mezirowian indicators of transformation.

3.3.1 Research Design and Rationale

A qualitative self-study was selected because it privileges practitioner knowledge and supports iterative changes. The researcher's position as an insider provides access to detailed information about classroom practices, which enables the study of teaching methods and their direct effects on teacher and student learning (Hawley, 2010; Lassonde & Galman, 2019). The researcher employed self-study techniques in conjunction with dialogic enquiry, involving TAU group members and critical friends, to achieve multiple viewpoints and minimise the influence of a single perspective.

This design is appropriate for the twin aims of (a) tracing how intentionally enacted character-strengths strategies generate reflective sequences in practice and (b) judging whether those sequences constitute instrumental adjustments or perspective transformations for educators and learners.

3.3.2 Context and Participants

The study took place within the TAU programme, where the lead researcher, who was also a participant, taught a scheduled cohort of 103 first-year students in Human Resource Management 1. All cohort members were invited to participate; participation was voluntary and subject to informed consent from the participants. The research team also included critical friends (colleagues experienced in reflective practice research) who provided ongoing peer debriefing and independent classroom observations. Students' participation was anonymised and separated from the summative assessment to preserve ethical integrity.

3.3.3 Data Sources and Collection Procedures

The study used multiple complementary qualitative artefacts across enactment–reflection cycles to enable triangulation and depth of data. Researcher sources included Gibbs-patterned reflective journals, annotated lesson plans that specified targeted character strengths, and selected audio-visual segments for stimulating recall. Student artefacts comprised anonymised reflective logs, short essays on character-strength activities, and course project reports that evidenced the enactment of the principles. Dialogic data consisted of semi-structured student questions, stimulated recall interviews with educators, and post-session debriefs with critical friends. Observational and documentary sources included field notes, programme materials, and pre- and post-reflective surveys with specific items. Data collection followed an iterative cycle of teaching, reflecting, collecting, interrogating, adjusting, and re-implementing, aligned with action research.

3.3.4 Researcher Positionality and Reflexivity

The researcher's dual role as an educator and investigator was explicitly acknowledged and treated as a source of insight rather than a bias to be eliminated. Reflexive practice was systematic; the researcher maintained meta-reflexive memos documenting assumptions, emotional responses, interpretive choices, and conflicts that emerged during the analysis (Jamieson et al., 2023). Critical friends reviewed reflexive memos and preliminary interpretations to identify alternative readings and reduce the likelihood of confirmatory bias. This reflexivity is integral to credibility in self-study, where subjectivity is harnessed, interrogated, and documented (Lassonde & Galman, 2019).

3.3.5 Analytic Framework: Integrating Gibbs' Reflective Stages with Mezirow's Transformative Indicators

The analytic strategy intentionally integrated Gibbs' stepwise reflective structure (description → feelings → evaluation → analysis → conclusion → action plan) as presented below (Fig. 3.2), the elicitation and micro-coding scaffold, with Mezirow's TLT providing the macro-interpretive criteria for identifying genuine perspective transformation (Aubrey et al., 2015; Christie et al., 2015; Taylor, 2000) (Fig. 3.3). Contemporary studies that utilise Gibbs to surface reflective moves and Mezirow to interpret depth informed the approach (Hashim et al., 2023; Ocampo et al., 2025). In practice, the two frameworks functioned together: Gibbs structures the reflective data so that sequences are visible, and Mezirow's concepts (disorienting dilemma, critical reflection on premises, communicative reflection, and enacted transformation) guide judgement about whether reflective sequences amount to instrumental learning or to perspective change.

3.3.6 Operational Coding and Stages

The analysis used an iterative, theory-informed protocol aligning Gibbs's reflective micro-processes with Mezirowian transformational indicators across multiple sources. Familiarisation and descriptive coding established factual episode elements and flagged disorienting dilemmas where practice conflicted with expectations or values. Affective and evaluative coding captured emotions—surprise, discomfort, and moral concern—recognised as catalysts for deeper reflection. Conclusion and action mapping tracked intended and enacted changes, treating sustained implementation of redesigned practices as evidence of embodied instrumental shifts. Cross-actor reciprocity mapping linked teacher actions to disorienting student experiences, subsequent critical student reflection, altered participation, and reciprocal

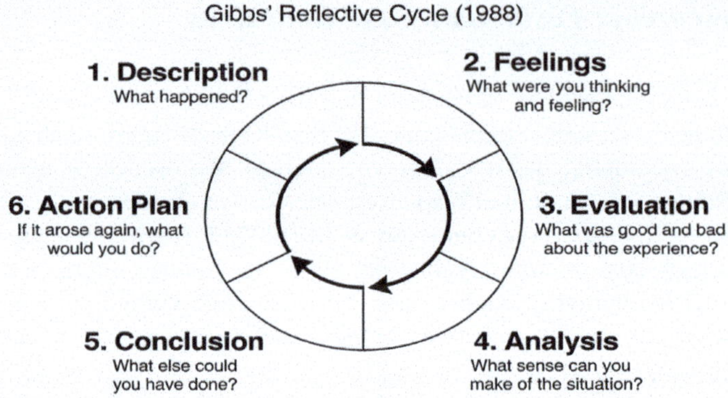

Fig. 3.2 Gibbs reflective cycle. *Source* authors' own, with assistance of perplexity AI

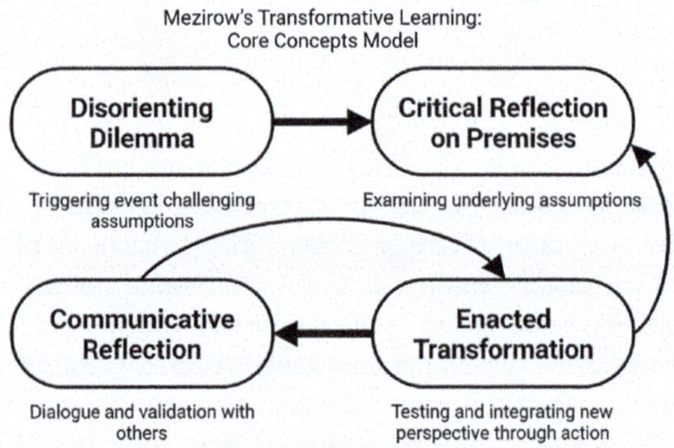

Fig. 3.3 Mezirow transformation learning theory. *Source* authors' own, with assistance of perplexity AI

teacher adaptation. Together, these stages traced how emotional and cognitive disturbances catalysed communicative reflection and practical transformation, while the protocol supported systematic identification, coding, triangulation, and claims about transformative learning across various educational settings.

3.3.7 Coding Procedures and Analytic Rigour

A codebook was developed that explicitly documented how Gibbs' stages corresponded to Mezirowian constructs (e.g., Gibbs' "feelings" → flags for potential critical reflection; Gibbs' "analysis" → evidence of interrogation of assumptions). The codebook contained definitions, exemplar quotations and decision rules to enhance consistency and reliability.

Member-checking provisional interpretations and theme maps were shared with participants (teacher and volunteer students) and critical friends for validation and to surface alternative readings (Lassonde & Galman, 2019).

Reflexive auditing: The researcher and critical friends maintained an audit trail (including raw data references, coding iterations, and memo chronology) to support the dependability and confirmability.

3.3.8 *Distinguishing Instrumental Versus Transformative Learning*

A central analytic task was to differentiate between instrumental changes (practical modifications without altered meaning perspectives) and transformative learning (reframing assumptions or identity). Operational criteria included: (i) Instrumental: changes limited to technique or procedure without language indicating changed beliefs about teaching, learning, or role identity. (ii). Transformative: language indicating reinterpreted beliefs (for example, "I view assessment as part of the learning process", "I see myself as a facilitator, not as a lecturer"), sustained enactment across subsequent cycles, and evidence of communicative testing (discussion with others that leads to revised practice) (Aubrey et al., 2015; Taylor, 2000). Where possible, changes were corroborated by multiple sources (teacher journals, student reflections, and observed classroom behaviour).

3.3.9 *Trustworthiness, Ethics and Limitations*

Rigour was pursued through triangulation (multiple artefact types), peer debriefing (critical friends), member checking, and maintenance of a clear audit trail (Hawley, 2010). Ethical procedures included institutional ethics approval, informed consent, anonymisation, secure data storage, and explicit separation of research participation from course assessment. The methodological limitations are those typical of self-study: the findings are richly contextual and not statistically generalisable but are transferable through thick description and transparent analytic accounting (Lassonde & Galman, 2019). Reflexivity and critical friend challenges were used to mitigate the risks of researcher centrism (Jamieson et al., 2023).

3.3.10 *Interpretive Integration and Reporting*

The reporting strategy mirrors the integrated analytic logic: the findings are presented in two interwoven layers. First, process traces organised by Gibbs stages show how specific classroom episodes unfolded and how reflective cycles progressed into action (micro-analytic layer). Second, interpretive accounts apply Mezirow's criteria to these traces to argue whether shifts represent surface, instrumental learning, or a more profound perspective transformation (macro-interpretive layer). Reciprocal growth is demonstrated when process traces for teachers and students converge, and when Mezirowian indicators of reframing and enacted practice appear on both sides of the pedagogical relationship. Throughout the reporting, exemplar quotations, short sequence vignettes, and reflexive memos were used to make analytic moves transparent and auditable.

3.4 Findings

Data analysis identified six key themes, three relating primarily to educator development (teacher identity and reflective practice, pedagogical innovation and instructional redesign, and leadership, agency, and social justice advocacy) and three to student development (strengths-based self-awareness and agency, students' collaboration and interpersonal growth, and academic engagement, resilience, and professional readiness).

Each theme captures a distinct dimension of growth catalysed by the Strengths Gym intervention and associated TAU activities. The subthemes detail the specific manifestations of these changes as observed in the dataset. The themes and their operational definitions are presented and analysed in the tables below. Table 3.1 focuses on educator development and Table 3.2 presents student development, with each definition clarifying the conceptualisation employed in this study.

Theme 3.1: Teacher Identity and Reflective Practice

Participation in the TAU programme and facilitation of *Strengths Gym* precipitated a substantial reconfiguration of the educators' professional identities, driven by iterative enactment–reflection cycles. Structured strength-based activities exposed classroom behaviours and student perspectives that contradicted the educator's prior teaching assumptions, creating disorienting dilemmas that initiated perspective shifts in students and prompted the educator's deep reflection. The educator's journals and field notes mapped neatly onto Gibbs' reflective cycle—describing events, expressing feelings, making evaluative judgements, analysing causes, drawing conclusions, and devising action plans—thereby scaffolding concrete pedagogical change. Exposure to VIA character strengths language and the *Strengths Gym* intervention provided practical tools, including weekly strengths intentions, structured reflection prompts, and classroom norms, which elevated the student voice. As a result, the educator transitioned from content transmitter to facilitator, prioritising student abilities, self-determination, and coping skills, and shifting from answer provider to strength-based supporter. This change was procedural, affective, and ethical, influenced by social justice scholarship within TAU, reinforcing inclusive facilitation as central to professional purpose. The process was reciprocal: student behaviours generated

Table 3.1 Fostering educator development through the strength-based intervention: themes, sub-themes, and operational definitions of the themes

Themes	Sub-themes	Operational definitions
Teacher identity and reflective practice	• Self-awareness and reflexivity • Reframing professional purpose • Ongoing reflective cycles	Educators' ongoing process of self-examination, professional self-concept formation and use of reflective inquiry
Pedagogical innovation and instructional redesign	• Educational technology and method diversification • Integration of strengths-based frameworks • Scaffolding reflective exercises	Deliberate adaptation of teaching strategies informed by reflective insights
Leadership, agency and social justice advocacy	• Inclusive leadership emergence • Influence and stakeholder engagement • Social justice pedagogy adoption	Educator's emerging leadership identity, empowerment to influence change and commitment to equity within the context of education

Table 3.2 Fostering student development through the strength-based intervention: themes, sub-themes, and operational definitions of the themes

Themes	Sub-themes	Operational definitions
Strengths-based self-awareness and agency	• Identification and articulation of personal strengths • Strengths in application to tasks • Increased confidence and ownership	Conscious identification and application of personal character strength by students
Students' collaboration and interpersonal growth	• Recognition of complementary team strengths • Trust, communication and interdependence • Problem-solving and group dynamics	Students' demonstration of teamwork, communication, and mutual respect during the learning process
Academic engagement, resilience and professional readiness	• Connection to academic goals and scaffolding needs • Career-oriented competencies and adaptability • Retention and motivation indicators	Students increased their motivation, persistence, and preparedness for professional challenges

questions, prompted small pedagogical experiments, and informed revisions, illustrating a Reciprocal Growth Model where student strengths in learning and teamwork enabled teacher innovation, which in turn fostered student development. Mechanistically, three linked components supported simultaneous educator and student growth:

data-driven reflexivity legitimised reflection through measurable outcomes; actionable language and tools made facilitation practicable; and repeated Gibbs-style cycles embedded changes into routine practice. In contexts such as South African universities, where many students enter underprepared, this facilitative, strengths-based approach creates more equitable learning opportunities and supports learners from disadvantaged backgrounds. Overall, strengths-based interventions catalysed teacher reflection, reframed their professional identity, and established ongoing reflective cycles. The resulting loop—students exhibiting strengths, the teacher reflecting and redesigning, and students developing further—describes how *Strengths Gym* within TAU fosters sustained reciprocal growth for both educators and students.

Theme 3.2: Pedagogical Innovation and Instructional Redesign

The implementation of the *Strengths Gym* intervention catalysed substantial pedagogical change, reshaping teaching methods, curriculum design, and classroom activities to prioritise student growth. The VIA framework and positive psychology constructs provide a shared vocabulary and conceptual scaffold. Hence, tasks such as strength identification and reflective follow-ups became core learning activities that shaped students' approaches and assessment criteria for their peers. This reorientation enabled the educator to design assignments that foregrounded strength development rather than merely correct answers, aligning with calls to embed fundamental pedagogical principles rather than surface-level innovations. Student engagement, however, varied: while 93% reported increased self-awareness, qualitative notes and observations showed an uneven ability to connect strengths with work in terms of discipline-specific goals. In response, the educator iteratively strengthened scaffolding explicit prompts that linked the named strengths to the course objectives. These rubrics rewarded strengths-based collaboration and Gibbs-aligned reflective templates, which made the pathway from reflection to academic application clearer.

A further strand of change involves methodological diversification and the purposeful use of educational technology. Digital tools for strengths mapping, peer feedback, and asynchronous reflection extended participation, particularly for reticent students; online charts and discussion forums supported out-of-class development and evidence submission. These blended methods adhere to best practices for adaptable learning spaces, allowing educators to combine basic and advanced facilitation techniques.

Collectively, these innovations demonstrate how strength-based interventions enable teachers to create instructional methods that support concurrent teacher development and student achievement. The VIA language and *Strengths Gym* activities provided actionable tools for designing strength-focused tasks, scaffolding converted reflective practice into meaningful academic work, and technological and methodological variation increased accessibility. The redesign process followed a Reciprocal Growth dynamic—student behaviours guided teacher modifications, producing measurable student progress that spurred further redesign. Practical implications include intentionally aligning strengths activities with outcomes and assessments, providing Gibbs-based reflection templates for students unfamiliar with reflexive work, investing in professional learning to translate VIA concepts into tasks, and ensuring institutional support (time, technology, and collaborative design) to sustain these changes, particularly in contexts where students enter underprepared.

Theme 3.3: Leadership, Agency and Social Justice Advocacy

The TAU programme catalysed an emergent form of inclusive leadership among educators that was both pedagogical and political. As the educator considered the communal learning experience, taking on supervisory roles within TAU became less about formal position-holding and more about demonstrating classroom behaviours that highlight dignity, voice, and shared respect. This enactment of leadership was evidenced in concrete actions: redesigning group tasks to surface complementary strengths, deliberately inviting quieter or first-generation students into roles of responsibility, and publicly advocating for assessment practices that recognised collaborative and strength-based contributions. The educator's narrative indicates that she found herself stepping up to convene discussions about inclusion and assessment changes because of her exposure to the *Strengths Gym* intervention. The intervention thus enabled her to envision equitable practice could look like. This confession illustrates how facilitation practice translates into the educator's ability to teach with purpose and clear goals in mind, as opposed to passively reacting to what is happening during a learning event. These findings align with scholarship that views innovation pedagogy as inherently connected to new forms of professionalism, where educators shape learning environments responsively (Halder, 2023; Sasson et al., 2022).

The second closely interwoven outcome was the adoption of social justice pedagogy. Thus, exposure to ideas within the TAU, particularly those associated with Bali's framing of pedagogical equity, sharpened the educators' sensitivity to systemic barriers and repositioned classroom design as a site of social advocacy. Strengths Gym provided an operational vehicle for this shift: VIA language and strengths-based activities made it possible to surface student assets rather than deficits and to restructure interactions so that marginalised students were afforded authentic participation in the process. Describing this transition as both ethical and practical, the educator recalls requesting that students map their strengths and set intentions, revealing that the instruction was designed to accommodate both less talkative and more vocal students. The educator based his choice on reflective-identity research, which demonstrates that critical self-reflection leads to ethical changes in teaching practices (Sahling & De Carvalho, 2021; Yazan, 2022). The approach follows innovation pedagogy principles, which require designers to create solutions that serve both equity and efficacy needs (Halder, 2023; Ramya, 2023).

The intervention implemented a third component, which extended its impact to areas outside academic learning spaces. The educator's narratives demonstrate that she applied her newly acquired skills and knowledge to discuss and promote student-centred approaches on various platforms. The educator not only discussed the *Strengths Gym* activities with other participants in the TAU programme and colleagues in the department but also presented this paper as a work in progress at the Vaal University Teaching and Learning Conference in October 2025. Conversations with Colleagues, informal demonstrations of *Strengths Gym* activities, and conference presentations to the university audience potentially created openings for integrating strengths-based, socially conscious practices across different modules in the university. Furthermore, she developed leadership skills through student progress observation and reflective practice, which enabled her to establish a small-scale initiative to test new approaches. The process of grassroots influence follows established models of pedagogical transformation, which depend on practitioners to spread new learning spaces (Sasson et al., 2022).

Analytically, these three subthemes operate together to answer the study's central question regarding how a strengths-based intervention can foster simultaneous educator and student development in a South African university context. *Strengths Gym* produced visible student enactments of capability and engagement; these enactments functioned as both moral prompts and practical evidence that motivated the educator to assume inclusive leadership, adopt social-justice pedagogical commitments, and mobilise colleagues. In turn, the educator's leadership and advocacy reshaped classroom structures and institutional conversations, creating further opportunities for student empowerment. Indeed, this mutual relationship between student development and teacher initiative creates a path which connects strength-based practice to teacher autonomy to promote student equality in educational settings where first-year students face multiple challenges.

The question that now arises is: What are the Implications of these findings for teaching and institutional practice? It is vital to take cognisance of the fact that a combination of professional development programmes with strength-based tools (for example, VIA vocabulary, *Strengths Gym* activities) and social-justice pedagogical framing can be utilised to enhance leadership inclinations that lead to inclusive classroom practice. Second, institutions should recognise and support educator-led innovation by providing time and platforms for sharing practices, as grassroots diffusion is a key mechanism for broader uptake. Third, embedding strengths-based tasks into assessment and curriculum design is necessary so that programme-level structures reinforce leadership at the classroom level; otherwise, inclusive experiments risk remaining isolated. These recommendations resonate with calls to design innovative and equitable new learning environments (Halder, 2023; Sasson et al., 2022).

The *Strengths Gym* intervention enabled the primary researcher to develop leadership skills through strength-based teaching methods, which, when combined with social justice principles, created teachers who could advocate for educational equity. This combination of practice and advocacy created a reinforcing loop in which student development validated teacher agency, and teacher agency expanded student access and voice, thereby demonstrating a viable route for simultaneous educator and student growth in South African higher education. To complement these findings, Table 3.2 focuses on student development, and the operational definitions clarify how these themes are conceptualised in this research.

Theme 1: Strengths-Based Self-awareness and Agency

Regarding student development, the implementation of *Strengths Gym* within the TAU programme produced clear evidence that a structured, strengths-based intervention cultivates student self-awareness and a growing sense of agency in the students. Quantitative feedback indicated that 93% of the participating students reported improved self-awareness, 87% reported improved relationships, and 85% agreed to continue developing their strengths independently after the intervention. This broad change was corroborated in reflective journals and classroom observations, where students were able to name specific character strengths and describe how those strengths influenced their actions in learning situations. This process of discovering and sharing individual strengths produced results that align with research findings, which show that students develop reflective abilities and form connections when they identify their personal strengths (Moldenhauer et al., 2021).

In the current study, students actively identified qualities such as perseverance, cooperation, curiosity, and leadership during group projects and tests. This serves as both a cognitive anchor and an affective resource. For instance, over 90% of students reported that, following the intervention, discovering their strengths helped them identify traits they had never realised they possessed. This shift was expressed in statements such as, “Helped me realise that I am capable of many things, I used to doubt myself but it turns out that I am really an amazing, creative and smart person,” and, “Before I was shy to make friends but I decided to give it a try and I currently have a couple of friends.” Several students described how recognition of their strengths expanded their willingness to participate: “I now know where my strengths lie. It has helped me understand we are different and have different strengths.” Such evidence demonstrates how the use of explicit strengths language both informed student choices and opened new pathways for experimentation and collaboration in the academic context.

Through the *Strengths Gym* intervention, students were guided to identify, articulate, and reflect on their character strengths in specific academic and collaborative contexts. Structured weekly activities invited students to set intentions for utilising a selected strength, such as curiosity, creativity, or a love of learning in classroom tasks or group assignments. Reflections captured in follow-up surveys and student work consistently showed that students became increasingly aware of how deliberate strengths use influenced their participation and approach to challenges in the course. For example, students described how applying identified strengths in group settings enabled them to communicate more confidently, collaborate effectively, and better appreciate the complementary abilities of their peers: “Now I have learnt to listen to their opinions no matter what”; “It has helped me to understand my unique character strengths… and I now know where my strengths lie”; “As a collective we can combine our strengths in order to do something productive.” These iterative activities and reflection prompts fostered a shift from passive self-identification to active experimentation with strengths in real academic tasks, thereby supporting both personal and social development. Rather than remaining an isolated wellbeing initiative, the strengths-based approach was woven into subject learning, contributing to enhanced engagement and purposeful academic participation in the subject.

Students’ increased confidence in their ability to participate, communicate, and take charge of both academic and group projects was a clear indication of their growing agency. Several students discussed how identifying and deliberately utilising their character strengths helped them make new friends, participate more freely in group discussions, and appreciate the distinctive qualities of themselves and their peers.

For example, one participant reflected, "Before I was shy to make friends, but I decided to give it a try and currently have a couple of friends." Another stated, "It has helped me utilise my character strengths to enhance interpersonal relationships." Students attributed their willingness to listen to others, take initiative, and contribute actively to teamwork directly to the strengths-based approach: "I now know where my strengths lie … as a collective we can combine our strengths to do something productive." While leadership and coordination did not always emerge formally, qualitative feedback consistently indicated that strength recognition nurtured self-efficacy and more active peer engagement. Importantly, these visible shifts in student behaviour became catalysts for the educator's own reflective practice, feeding the Reciprocal Growth Model central to this study. Observing students applying their strengths inspired further redesign of classroom activities to support autonomy, fostering an ongoing cycle of mutual development between students and educators.

Together, these findings address the central research question by demonstrating that the *Strengths Gym* fosters student self-understanding and agency in ways that are reciprocally linked to teacher practice. When students can identify, apply, and reflect on their strengths, they not only improve their task performance and participation but also generate an empirical and moral impetus for educators to adapt their pedagogy to facilitate and promote equity. Therefore, the implications for teaching are concrete: instructors should integrate explicit strengths-language and regular, scaffolded reflection into course activities, align strengths tasks with assessment criteria to enhance perceived relevance, and provide simple self-monitoring tools (e.g. weekly intention logs) to sustain practice—strategies that are supported by both strengths- and innovation-pedagogy literature (Halder, 2023; Ramya, 2023; Sasson et al., 2022). Additionally, creating structured opportunities for students to assume leadership roles within tasks can amplify both individual agency and group functioning, a benefit particularly salient in contexts with a high proportion of first-generation students.

Strengths Gym fostered a cluster of student outcomes, including more apparent self-awareness, pragmatic application of strengths to learning tasks, and enhanced confidence and ownership, which together constitute a developmental pathway. Student achievements create a dynamic relationship between teacher reflection and instructional redesign, illustrating how the strengths-based intervention in TAU leads to simultaneous growth in both teachers and students in South African universities.

Theme 2: Collaborative Competence and Interpersonal Growth

This theme illustrates how students modified their teamwork methods, social interactions, and problem-solving skills in response to the interventions. Across class observations, reflective journals, and post-intervention feedback, students reported and demonstrated a growing capacity to recognise complementary abilities within teams, build trust and interdependence, and manage group processes more productively. These interpersonal developments did not occur as incidental byproducts; they emerged directly from the structured strengths activities that foregrounded assets rather than deficits, providing clear, observable markers that teachers could use to refine their facilitation in real time. Through activities that required them to identify and share their VIA character strengths, groups discovered different skills and allocated roles among themselves according to their skill sets, with greater intentionality.

Students transitioned from task distribution based on convenience or hierarchy to assignments grounded in identified strengths, where one student described themselves as strong in organisation, another claimed creativity, and teams deliberately combined these orientations to plan and execute their projects. One student reflected, "I realised my teammate's creativity balanced my planning—together our work improved," a sentiment echoed in multiple reflective entries. Previous studies have confirmed that this pattern exists in other strengths-based curricula which use direct strengths identification to enhance relationship development and leadership abilities (Moldenhauer et al., 2021). In the TAU context, the practical consequence was improved team cohesion and a more strategic use of members' capacities during complex tasks.

Concomitant with the recognition of complementary strengths was a notable increase in trust, communication, and interdependence, which were particularly salient for first-generation students. *Strengths Gym's* routine of weekly intentions and follow-up reflections created predictable, low-stakes opportunities for students to disclose their capabilities and receive affirming peer feedback, which reduced social risk and increased participation over time. Observational data, combined with student feedback, revealed that students modified their communication style by showing respect to one another, making space for quiet members, and utilising the expertise of team members when tackling new tasks. These findings support innovation pedagogy principles, which focus on creating learning spaces that promote fair opportunities for all students (Halder, 2023; Sasson et al., 2022).

For first-year students, who described many early experiences of classroom marginalisation, the structured strengths approach constituted a first meaningful experience of purposeful teamwork grounded in mutual respect, increasing their sense of belonging and willingness to contribute.

The third interrelated development comprised improvements in group problem-solving and overall dynamics. Teams that adopted strengths-aware practices moved more quickly from impasse to solution; they used strengths maps to reframe challenges (for example, treating a conflict as an opportunity to apply perspective-taking or fairness) and to distribute cognitive labour in ways that maximised efficiency and creativity. In practice, this meant more effective role negotiation, fewer prolonged disputes, and more integrative solutions to assignment problems in the future.

These shifts align with research evidence indicating that reflective tasks incorporating strengths-based approaches enhance team problem-solving abilities and leadership skills among team members (Moldenhauer et al., 2021; Ramya, 2023). Simultaneously, the analysed data showed that the groups experienced difficulties when they faced time constraints and insufficient experience. Thus, when time pressure and limited familiarity with reflective work persisted, some groups defaulted to superficial collaboration. This unevenness suggests that while strength approaches can catalyse improved group functioning, they require deliberate scaffolding to translate interpersonal gains into consistent academic performance.

These findings demonstrate that students develop essential social abilities through strengths-based collaboration, which teachers can use to make educational adjustments. The Reciprocal Growth Model becomes apparent in this situation, as students' improved teamwork skills and their noticeable strengths lead teachers to establish new support systems. The teacher further changes and assigns task responsibilities according to a student's strength profile. The evaluation methods can also be modified accordingly. The teacher's reflection on the observed dynamics, captured through Gibbs-style cycles, led to further instructional changes that reinforced student growth, demonstrating how student and educator development proceed in tandem.

The implications for teaching derived from these findings include the need to embed explicit strengths-mapping activities into group task design, provide short, structured reflection templates that help students link strengths to task outcomes, and align assessment rubrics with collaborative and strengths-based behaviours so that interpersonal competence is formally valued. Professional development that trains educators to read and respond to strengths-based interactions will strengthen the reciprocal loop and prevent superficial adoption. Providing institutional support through time allocations for scaffolded group activities and basic digital tools for strength mapping and peer feedback tracking can maintain the achieved gains (Halder, 2023; Sasson et al., 2022).

Strengths Gym's advanced collaborative competence and interpersonal growth enable students to recognise complementary strengths, develop trust and interdependence, especially among first-generation learners, and improve group problem solving. These student developments, observable and interpretable by the educator, created the conditions for reciprocal pedagogical refinement and the concurrent growth of both students and teachers within the TAU programme.

Theme 3: Academic Engagement, Resilience and Professional Readiness

The *Strengths Gym* intervention yielded notable improvements in students' resilience, academic engagement, and professional readiness, with the most pronounced results observed when robust instructional scaffolding was in place. This multifaceted effect is supported by both quantitative and qualitative data, which demonstrate that students exhibit a greater ability to actively engage with academic material and utilise their personal qualities to navigate group dynamics and assignments. As one participant reflected, "Me knowing my strengths can make me a better version of myself and thrive more in life." Reports of increased confidence and agency are further corroborated by students who noted, for example, "Before I was shy to make friends, but I decided to give it a try and currently have a couple of friends" and "It has helped me to communicate more. I now understand myself, so it is easier for me to understand other people". These findings are consistent with established literature demonstrating that strengths-based interventions and collaborative group activities are effective mechanisms for fostering self-efficacy, motivation, and adaptive leadership in higher education contexts (Moldenhauer et al., 2021) Although variability in student gains remains evident, the breadth and consistency of self-reported outcomes underscore the value of character strengths pedagogies for advancing both individual and collective development.

Ninety-five per cent of respondents acknowledged the importance of character characteristics in their future positions as human resource professionals. It was emphasised how character qualities help students prepare for professional HR work. Students discussed how they were able to see themselves as more competent and flexible workers because they discovered and utilised their talents. As one participant shared, "Me knowing my strengths can make me a better version of myself and thrive more in life," while another explained, "It helps me to compile a list of professional strengths that can be used for professional roles." The connections between self-awareness and effective teamwork, crucial for HR roles, were also evident "If you are aware of your strengths it is easier to identify the strengths of others making it easier for you to work with them as you will understand one another's qualities" Other responses explicitly linked strengths work to overcoming professional challenges, with one student noting, "You will face different difficult challenges as an HR Professional." Together, these reflections confirm that the intervention not only fostered growth in individual well-being and relational skills but also equipped students with a mindset and repertoire relevant to professional effectiveness in the HR field.

Although data analysis revealed positive trends, as discussed in the preceding paragraphs, classroom observations, along with some student stories, demonstrated that students had varying abilities to utilise reflective exercises to achieve academic goals in the module. Some Students demonstrated different abilities to link their strengths-based reflections to their exam studies, teamwork responsibilities, and project development tasks. The observed differences between students align with research findings on strength measurement, which indicate that regular self-checks benefit certain students. In contrast, others require additional support to understand their monitoring activities (Krentzman & Gass, 2024). In this study, which involved all first-year students, the capacity to bridge reflective insight and academic practice depended on the clarity of instructor prompts, the availability of short reflective templates, and explicit discussions of how a named strength could be applied to meet a specific course outcome.

When scaffolds were deliberately provided, such as rubrics that rewarded evidence of strengths use, short Gibbs-style reflection prompts, and instructor modelling of how strengths map onto assessment tasks, students demonstrated improved application of strengths to academic work and greater resilience in the face of setbacks. These competencies included improved role negotiation in group projects, improved time management linked to prudence and perseverance, and more adaptive approaches to ambiguous tasks. These emerging educational approaches enable students to develop the career-ready competencies they require. Students not only learn to work together more effectively, but they also learn how to solve problems and consider different perspectives on the same issue. These skills are vital for achieving success in today's workplace. The research results confirm innovation-pedagogy theories, which support learning spaces that teach adaptable competencies through multiple teaching methods, thereby preparing students for platform-based work challenges (Halder, 2023; Ramya, 2023; Sasson et al., 2022).

The data revealed multiple indicators of what would happen to students motivation and retention rates. The *Strengths Gym* programme enabled students who previously stayed away from classes due to work and family responsibilities to maintain their academic involvement. The programme established a new method for students to build feelings of belonging and competence through their work on tasks. Academic work became more accessible to students through their contributions, which simultaneously decreased their feelings of academic isolation. The research demonstrated that well-designed strength-based interventions serve as protective measures, helping students remain in university by increasing their interest in their work and enhancing their understanding of its value. Additionally, it also showed that motivational gains from these interventions will not continue unless teachers establish direct connections between programmes, assessment systems, and provide ongoing support.

Pedagogically, these findings imply several concrete adjustments are needed. First, implementing strength-based reflection requires structured support to convert reflective thoughts into academic work. This could be done through assessment-based reflective tasks, project examples, and feedback that focuses on the application of strengths. The integration of workplace scenarios into strengths activities by instructors can enhance career readiness connections through team simulation tasks, role-playing activities, and employer-developed briefs that demonstrate how students can apply their competencies. However, it is important that the intervention design is equipped with built-in monitoring and measurement systems.

These systems would track student development through self-assessment checks, peer feedback systems, and digital portfolio tracking, which minimise measurement effects (Krentzman & Gass, 2024). Finally, the successful implementation of strengths-based practice at the institutional level requires three essential elements: time for redesign, access to technology, and recognition of strengths-based practice in workload models (Halder, 2023; Sasson et al., 2022).

The Reciprocal Growth Model serves as the primary link between teacher and student development throughout all identified themes. The development of student self-awareness, practical competencies, and collaboration skills prompted teachers to examine their teaching methods, resulting in the implementation of new approaches that support inclusive leadership and advocacy (Themes 1 and 2). This process requires purposeful support systems and precise assessment methods to create lasting academic achievements and professional growth. Indeed, in the context of this study, universities need to implement strength-based support systems which connect assessment methods and career-oriented tasks. These measures could help the reciprocal model maintain student resilience and facilitate teacher development.

3.5 Conclusion

This self-study demonstrates that intentionally enacted character-strengths interventions, scaffolded through structured reflective practice, can catalyse reciprocal growth for both educators and students within the TAU programme. By aligning Gibbs-style micro-reflection with Mezirowian indicators of transformation, the study traced how iterative enactment–reflection cycles produced both instrumental improvements (pedagogical adjustments and novel activities) and, in many cases, more profound perspective shifts (reframed teacher identity, strengthened student agency, and professional readiness). Reciprocity emerged when teacher actions elicited disorienting experiences and critical reflection in students, prompting educators to re-examine their practice and effect sustained change.

Practical implications flow across three, interdependent levels. For professional development, programmes should embed structured reflective frameworks (for example, Gibbs' Cycle), require implementation and critical evaluation of strengths-based interventions in authentic contexts, and cultivate sustained communities of practice to support ongoing refinement of the intervention. For curriculum design, character strength activities should be integrated alongside disciplinary content, with scaffolded reflective exercises that accommodate students new to reflective work and explicit links to employability and professional skills. At the institutional level, policies must protect time and resources for pedagogical experimentation, recognise reflective teaching and curriculum innovation in reward structures, and prioritise equity-focused interventions that address the psychosocial needs of first-generation and marginalised learners.

Finally, it is necessary to note that the limitations of the self-study design affect statistical generalisation, although they improve contextual transferability due to detailed descriptions and clear methodological explanations. Future researchers should investigate how strengths-based teaching methods impact teacher development and student achievement outcomes over time while testing the Reciprocal Growth Model in various academic fields and educational settings. Researchers should also conduct studies across multiple sites to determine how the model works at different scales and what specific conditions limit its effectiveness in terms of transformative reciprocal learning.

References

Advocates for International Development (A4ID). (2022). *SDG 4: Quality education—A legal guide*. London, United Kingdom: Advocates for International Development (A4ID). Retrieved from www.a4id.org

Aubrey, K., Riley, A., & Mezirow, J. (2015). Transformative learning. In *Understanding and using challenging educational theories* (3rd ed.), (pp. 319–340). Sage. https://doi.org/10.4135/9781036232627.n21

Bali, M. (2020). *Critical digital pedagogy: A definition*. Hybrid Pedagogy. https://hybridpedagogy.org/critical-digital-pedagogy-definition/

Christie, M., Carey, M., Robertson, A., & Grainger, P. (2015). Putting transformative learning theory into practice. *Australian Journal of Adult Learning, 55*(1), 9–30. https://api.semanticscholar.org/CorpusID:145182605

Clifton, D. O., & Nelson, P. (1992). *Soar with your strengths*. Delacorte Press.

Cranton, P. (2023). *Understanding and promoting transformative learning: A guide to theory and practice*. Routledge.

de Kadt, E. (2020). Promoting social justice in teaching and learning in higher education through professional development. *Teaching in Higher Education, 25*(7), 872–887. https://doi.org/10.1080/13562517.2019.1617685

Department of Education. (1997). *Education White Paper 3: A programme for the transformation of higher education*. Pretoria: Department of Education.

Department of Education. (2001). *Education White Paper 6: Special needs education: Building an inclusive education and training system*. Pretoria: Department of Education.

Fung, D. (2017). Strength-based scholarship and good education: The scholarship circle. *Innovations in Education and Teaching International, 54*(2), 101–110.

Gibbs, G. (1988). *Learning by doing: A guide to teaching and learning methods*. Oxford Polytechnic.

Halder, I. (2023). Innovation pedagogy of the 21st century: Challenges and opportunities. *International Journal of Multidisciplinary Research, 5*, 1–6. https://doi.org/10.36948/ijfmr.2023.v05i03.3339

Hashim, S. N. A., Yaacob, A., Suryani, I., Asraf, R. M., Bahador, Z., & Supian, N. (2023). Exploring the use of Gibbs' reflective model in enhancing in-service ESL teachers' reflective writing. *Arab World English Journal, 14*(2). https://doi.org/10.2139/ssrn.4497558

Jamieson, J., Govaart, G., & Pownall, M. (2023). Reflexivity in quantitative research: A rationale and beginner's guide. *Social and Personality Psychology Compass, 17*(4). https://doi.org/10.1111/spc3.12735

Kitchenham, A. (2008). The evolution of John Mezirow's transformative learning theory. *Journal of Transformative Education, 6*(2), 104–123. https://doi.org/10.1177/1541344608322678

Knowles, M. S. (1984). *The adult learner: A neglected species* (3rd ed.). Gulf Publishing.

Lassonde, C. A., & Galman, S. (2019). *Self-study research methodologies for teacher educators* (Vol. 7). Leiden, The Netherlands: Brill Sense

Leibowitz, B. (2017). *Higher education for the public good: Views from the South*. Trentham Books.

Linkins, M., Niemiec, R. M., Gillham, J., & Mayerson, D. (2015/2014 online version). Through the lens of strength: A framework for educating the heart. *The Journal of Positive Psychology, 10*(1), 64–68

Lopez, S. J., & Louis, M. C. (2009). The principles of strengths-based education. *Journal of College and Character, 10*(4), 1–8.

Mezirow, J. (1991). *Transformative dimensions of adult learning*. Jossey-Bass.

Modry-Mandell, K., & Nguyen, M. (2022). A strength-based approach for co-constructing strong pedagogical partnerships and inclusive and equitable classroom environments. *Teaching and Learning Together in Higher Education, 1*(37), 3.

Moldenhauer, L., Osland, M., & Waterbury, T. (2021). Study abroad strengths-based curriculum: Advancing self-reflection and relationship-building leadership skills. *Frontiers: The Interdisciplinary Journal of Study Abroad, 33*(2), 79–92. https://doi.org/10.36366/frontiers.v33i2.637

Ocampo, F. A., McIntyre, B. G., Massey, D., & Saunders, R. (2025). Nursing student reflections on a research internship: A reflective discussion following the Gibbs' reflective cycle. *Nurse Education Today, 145*, Article 106481. https://doi.org/10.1016/j.nedt.2024.106481

Pendlebury, S., & Enslin, P. (2004). Social justice and inclusion in education and politics: The South African case. *Journal of Education, 34*, 31–48.

Peterson, C., & Seligman, M. E. P. (2004). *Character strengths and virtues: A handbook and classification*. Oxford University Press.

Pinnegar, S., & Hamilton, M. L. (2009). *Self-study of practice as a genre of qualitative research: Theory, methodology, and practice*. Springer.

Proctor, C., & Fox Eades, J. (2019). *Strengths gym: Build and exercise your strengths!* Guernsey: PPRC.

Ramya, B. (2023). Innovative pedagogical approaches. *Formation of Competencies of Gifted Individuals in the System of Extracurricular and Higher Education* (2), 17–29. https://doi.org/10.18372/2786-823.1.18161

Ryan, M. (2011). Improving reflective writing in higher education: A social semiotic perspective. *Teaching in Higher Education, 16,* 99–111. https://doi.org/10.1080/13562517.2010.507311

Sahling, J., & De Carvalho, R. (2021). Understanding teacher identity as an international teacher: An autoethnographic approach to (developing) reflective practice. *Journal of Research in International Education, 20*(1), 33–49. https://doi.org/10.1177/14752409211005380

Samaras, A. P. (2011). *Self-study teacher research: Improving your practice through collaborative inquiry.* SAGE.

Sasson, I., Yehuda, I., Miedijensky, S., & Malkinson, N. (2022). Designing new learning environments: An innovative pedagogical perspective. *The Curriculum Journal, 33*(1), 61–81. https://doi.org/10.1002/curj.125

Schön, D. A. (1983). *The reflective practitioner: How professionals think in action.* Basic Books.

Soria, K. M., & Stebleton, M. J. (2012). First-generation students' academic engagement and retention. *Teaching in Higher Education, 17*(6), 673–685.

Taylor, E. (2000). Fostering Mezirow's transformative learning theory in the adult education classroom: A critical review. *Canadian Journal for the Study of Adult Education, 14*(2), 1–28. https://doi.org/10.56105/cjsae.v14i2.1929

Thomas, L. (2010). Student retention in higher education: The role of institutional habitus. *Journal of Education Policy, 17*(4), 423–442. https://doi.org/10.1080/02680930210140257

Tolich, M. (2010). A critique of current practice: Ten foundational guidelines for autoethnographers. *Qualitative Health Research, 20*(12), 1599–1610.

Weick, A., Rapp, C., Sullivan, W., & Kisthardt, W. (1989). A strengths perspective for social work practice. *Social Work, 34*(4), 350–354.

Wilcox, P., Winn, S., & Fryvie-Gauld, M. (2005). 'It was nothing to do with the university, it was just the people': The role of social support in the first-year experience of higher education. *Studies in Higher Education, 30*(6), 707–722. https://doi.org/10.1080/03075070500340036

Yazan, B. (2022). Reflective practice as identity work: A teacher educator's reflections on identity tensions. In Z. Tajeddin & A. Watanabe (Eds.), *Teacher reflection: Policies, practices and impacts* (pp. 164–178). Bristol, Blue Ridge Summit: Multilingual Matters. https://doi.org/10.21832/9781788921022-015

Yorke, M., & Thomas, L. (2003). Improving the retention of students from lower socio-economic groups. *Journal of Higher Education Policy and Management, 25*(1), 63–74. https://doi.org/10.1080/13600800305737

Chapter 4
The Significance of Using Hip Hop Pedagogy as a Teaching Method in Poetry at a Selected School in Gauteng

Lerato Jerome Mosebi

Abstract There are two major approaches to teaching English poetry: traditional and progressive. Teaching English poetry in South Africa and elsewhere often follows traditional text-based methods that expose learners to unidimensional ways of learning poetry. This creates a gap between the poetry taught at school and the poetry that learners experience in their social spaces. Against this background, this study investigated the methods and practices used in teaching English poetry to Grade 11 Home Language (HL) learners in a selected secondary school in Gauteng, with a focus on identifying weaknesses and opportunities for improvement in relation to culture. This study employed a qualitative interpretivist approach. In-depth interviews and direct observations were used as key tools to gather data from two English Home Language (HL) Grade 11 teachers and learners during poetry lessons. The findings from this study indicate that poetry teaching, as prescribed by the Curriculum and Assessment Policy (CAPS), does not resonate with learners' cultural backgrounds, nor does it meet their literary and emotional needs. It is recommended that educators use knowledge derived from learners' local cultures and social worlds by adopting approaches such as Hip Hop Pedagogy to create greater access to and enjoyment of English poetry. This aligns with SG4, where learners gain relevant skills, universal literacy, and vocational skills for decent jobs and entrepreneurship.

Keywords English poetry · Socio-cultural theory · Language-in-education policy · Curriculum and assessment policy statement (CAPS) · Hip hop pedagogy

---

L. J. Mosebi (✉)
Department of Tourism and Integrated Communication, Faculty of Human Sciences, Vaal University of Technology, Vanderbijlpark, South Africa
e-mail: leratomo@vut.ac.za

K. Dube and L. M. Tjabane (eds.), *Advancing Sustainable Development Goals in Higher Education*, https://doi.org/10.1007/978-3-032-27646-9_4

4.1 Introduction

This section focuses on teaching English poetry to Grade 11 English Home Language (HL) learners at a selected secondary school in Gauteng, South Africa. It started by providing a background, problem statement, and research questions and aims. The background section further reviews how poetry is traditionally taught in the United Kingdom, Nigeria, and South Africa. The methods used in teaching poetry in South Africa have not changed much since the first democratic elections in 1994, despite several transformational and incisive educational reforms. Despite poetry's evolution into many other types of representations, poetry lessons are still characterised by traditional teaching strategies that are restricted to text (Department of Basic Education [DBE], 2011a, 2011b). This creates a divide and exclusion between the poetry learned in school and the contemporary poetry that learners are exposed to in their local communities and on virtual media. For example, poetry learners' experiences in Black Entertainment Television (BET) include poetry slams, unplugged music festivals, and audio-visual poetry performances, which are different from the poetry taught in schools (Newfield & D'abdon, 2015).

Most learners appear to be more interested in listening to poetry recitations on YouTube, Spotify, and iTunes than the English poetry offered in school. This reveals an exclusion between the poetry taught in school and the poetry which learners' experience and enjoy in their everyday lives. Tapping into the learners' local culture in poetry teaching naturally creates a portal for a pedagogical construct loosely framed as "Hip Hop Pedagogy." Hip Hop Pedagogy aims to include tenets of hip-hop culture in education, such as rap, graffiti, deejaying, breakdancing, and performance poetry (Akom, 2009) This is quite the opposite of teaching poetry from a neo-conservative traditional approach, which linguistically raptures African learners into Western Culture. Instead, Hip Hop Pedagogy affirms learners' African cultural identity, making them better versions of who they already are at home. Hip Hop Pedagogy makes learners aware of their global, national and local surroundings, meaning it educationally immerses them into the real world. Mooney (2022) describes Hip Hop Pedagogy as a "site of radical healing and critical literacy, using hip-hop aesthetics." The learners gain the scientific side of poetry with cultural relevance.

However, since Hip Hop Pedagogy is broad and new, for the purpose of this chapter, the focus will be on variables that can be directly be applied in the poetry classroom like Transpoemation, DopeLearning, TwitSong, Google Sculpting, Transtextuality, and Blackout poetic Transcription. Hip Hop Pedagogy is important in rendering inclusive and quality education especially when it comes to the poetry teaching, because it draws on learners' poetry turf and makes its scientific elements accessible. Poetry should be easy to access because access to basic education is a human right (Constitution of the Republic of South Africa, 1996b). If poetry is not taught in a culturally relevant manner, schools are continuing the apartheid legacy in the post-apartheid era. Teaching poetry using Hip Hop Pedagogy also aligns with the Sustainable Development Goal 4 of UNESCO (UNESCO, 2025). This ensures inclusive, equitable, and quality poetry education relevant to the African context (UNESCO, 2025).

Against this background, the following problems arise. Teaching English poetry in South African schools is challenging, as most learners fail this section of the English examination. Even those learners who manage to obtain a higher mark in poetry lose interest in poetry because of its lack of relevance to the contemporary world. Most English teachers fail to use multiple approaches to teach poetry to meet learners' diverse learning styles (Cooper, 2020).

This chapter seeks to answer the following research questions:

- What are the current methods used to teach English poetry?
- How relevant are current English poetry teaching methods?
- How can Hip Hop Pedagogy be used to improve the teaching of English poetry?

4.2 Literature Review

4.2.1 Current Methods Used to Teach Poetry

In the United Kingdom (UK), poetry taught in schools is often viewed as strange because it is not related to learners' real lives (Xerri, 2014a, 2014b). In a study conducted by Lambirth et al. (2012) on the teaching of poetry, the results indicated that most teachers do not teach poetry for enjoyment, but they teach poetry to assess. According to Dymoke and McGuinn (2021), strategies for teaching poetry in the UK remain unchanged; as such, poetry lessons are framed with a focus on skills such as making inferences, analysing poetry, critiquing poetry, and the limited writing of traditional poetry. Their findings are further supported by Delchamps (2018), who stated that poetry is the least well-taught subject in all languages offered in schools. The country's school curriculum is compressed with traditional academic objectives, and as such, not enough time is allocated to teach poetry writing (Dymoke & McGuinn, 2021). The traditional approach to teaching poetry swims against the tide of the skills learners need to progress outside the classroom because schools are also social institutions (Dewey, 1938).

In Nigeria, English poetry taught in schools is viewed as foreign because it is not related to the learners' social and cultural life. This is because Nigeria is a British colony, similar to South Africa; as a result, the teaching methods are still not aligned with the African Renaissance, especially in the post-colonial era. They have been largely inherited from the UK and transferred to Nigeria and South Africa, respectively. In Nigeria, the enjoyment of poetry is considered a symbol of economic prosperity, and poetry is seen as an art that represents a sophisticated cosmopolitan life (Ugwuozor, 2020). As a result, learners develop an inferiority complex when it comes to poetry learning (Ugwuozor et al., 2020).

Previously, the Nigerian schooling system prescribed canonical poetry texts, which were officially assessed. At the moment, poetry teaching is disconnected from other parts of the curriculums such as drama and literature (Akinkurole et al., 2021). Poetry is taught without written examinations (Ugwuozor, 2020). Most Nigerian schools appear to be integrating the teaching of poetry with the teaching of other subjects such as mathematics. According to Maduabuchi and Maduagwu (2015) and Akinsola (2014), the linguistic aspects of mathematics can be taught effectively using poetry. The rationale for integrating poetry into the curriculum is to capture learners' attention. For example, in mathematics class, the use of poetry enables the summation of mathematical explanations.

Learning Maths through poetry is powerful and unique. Maduabuchi and Maduagwu's (2015) research confirms that this method of teaching mathematics yields better results than traditional teaching. Akinsola (2014) is of the opinion that some aspects of Maths require thinking patterns of a poet rather than of a Mathematician because they are more visual, language orientated and metaphorical. The integration of poetry and mathematics has also been criticised by those who prefer teaching poetry as a stand-alone subject because it may lose its specialised and disciplinary nature in schools (Okwor & Chidiebere, 2021). The integrated teaching of poetry has resulted in many rich families registering their children for private poetry lessons in the country (Akinkurolere et al., 2021).

To expand on the teaching of English poetry in South African schools, the Curriculum and Assessment Policy Statement (CAPS) Grades 10–12 English First Additional Language (EFAL) (DBE, 2011a) and the CAPS Grades 10–12 HL (DBE, 2011b) prescribe a text-based approach to teaching poetry based on pre-reading, reading, and post-reading activities. During the pre-reading phase, learners' prior knowledge is stimulated only as a formality to facilitate poetry learning. Thus, poetry is taught in a standardised manner in South African schools (Cooper, 2020).

Other methods of teaching poetry in South Africa include educators teaching poetry verses line by line and augmenting those verses with discussion (Cooper, 2020; Mavhiza & Prozesky, 2020). The main objective of teaching poetry lessons in South African classrooms is to understand poetry for traditional assessment purposes.

4.2.2 Hip Hop Pedagogy: A Progressive Poetry Teaching Approach

One of the tenets of Hip Hop Pedagogy is to teach poetry through a strategy introduced by Emert (2015) known as transpoemation. Transpoemation is also referred to as "mentor texts" (Flores, 2021:65; McFarland, 2020:2). First, teachers must present a lesson on a target pop culture poem and its poetic devices to students. The lessons must be structured in such a way that learners can tease out the poetic devices and then use these poetic devices and structures to write their own poems. For example, repetition, similes, metaphors, wordplay, line breaks, exaggerations, and other aspects of poems must be made explicit to learners to enable them to model in their own poetry writing (Emert, 2015). This pedagogic technique involves learners using the framework of poems written by established popular culture poets to create their own poems. Transpoemation can also be used to let learners model their verbal recitation techniques based on other poets (McFarland, 2020). Teachers can use digital poetry tools found on YouTube for learners to imitate and then request that they verbally recite their own poems using the target accent.

Other tenets of Hip Hop include DopeLearning, TwitSong, and Google Scalpting (Häkkinen, 2022; Lamb et al., 2017; Lawati & Rana, 2022). DopeLearning is a teaching method in which learners combine existing rap songs to create new songs/ poems (Margento & Murat, 2021; Nicolau, 2022). Learners can use the lyrics from their favourite rap songs and remix them using their own punchlines to create new poems/rap music (Lamb et al., 2017). Another aspect of the DopeLearning strategy is the rearrangement of the lyrics of two or more hip-hop artists based on similar themes (Häkkinen, 2022; Lamb et al., 2017; Lawati & Rana, 2022). For example, the "Dear Mama" Lyrics by Tupac Shakur can be remixed with lyrics from J. Cole's "Apparently" from the Forest Hill Album, as both albums are dedicated to the mothers of both rappers. Other songs dedicated to the mothers of rappers that can work well with the DopeLearning teaching technique include "Donda" by Kanye West and JayZee's "Smile" from the 4:44 album. These are examples of how DopeLearning is used to remix songs. Hip Hop Deejays, also use Dope Learning to remix songs that have a similar themes or tune to entertain their audience. Similar to scaffolding, when learners design their own poetry, they must let go of the framework they used to remix their poems, as they would have mastered the art of writing their own poetry.

Another method of teaching poetry is when learners extract online tweets and turn them into poems. This method of teaching poetry writing is called TwitSong (Häkkinen, 2022; Lamb et al., 2017; Lawati & Rana, 2022). The TwitSong method of teaching poetry writing involves extracting knowledge on Twitter on a target topic (Häkkinen, 2022; Lamb et al., 2017; Lawati & Rana, 2022). Target knowledge becomes the basis for poems written by learners. Learners extract poetic devices from online tweets for educational purposes as they are relatable, or they can rewrite the tweets using structures like limericks, sonnets, or epitaphs. Twitter (now called X) is a good tool to use because it largely corresponds with poetry, as it is radical and has a wealth of literary knowledge.

Poetry can also be taught using Google Sculpting. Google sculpting encourages learners to search for popular terms, punchlines, and phrases from the internet and then convert them into classic poems. Rhythm and Poetry (RAP) music (Häkkinen, 2022; Lamb et al., 2017; Lawati & Rana, 2022). It is said that Google sculpting can enable the creation of poetic knowledge that transcends generations, as it is derived from classic English phrases.

Trans-textuality, another tenet of Hip Hop Pedagogy, is defined by Creely (2019) as a writing technique in which a text evokes another unrelated text, prompting the writer to integrate the two unrelated texts to create new knowledge. This strategy is usually used in literature studies or research where connections between unrelated texts are formed to discover new knowledge (Creely, 2019). However, the researcher would like to reframe this theory into poetry pedagogical practice, where a poem triggers unrelated art forms such as fashion, painting, drama, literature, dance, and building sculptures. Machado et al. (2017) mention that teachers need not test learners' understanding of poetry through text alone, but they could request them to dress up in the selected RAP theme to assess their creativity and understanding. Kanye West is an example of a hip-hop artist who regularly translates his hip-hop music into fashion or sneakers. One can easily see a Kanye West fan when they interpret their style of dress, because Kanye West applies transtextuality, where his music is linked to his own style of dress. In contemporary South African Hip-Hop music, Nadia Nakai (2019), a prominent female South African rapper, constructed a documentary music show reciting rhythm and poetry (RAP) on BET using transtextuality. She recited hip-hop music and made a sculpture of herself, painted it as she was storytelling and teaching about her personal life. The music documentary was called "Nadia Naked." This technique can help teachers link poetry which learners may not like, with other art forms that they love.

Blackout poetic transcription occurs when learners take a prose text they like and blackout all non-poetic aspects of the text with a marker (Keith & Endsley, 2020). This teaching strategy can assist in developing learners' skills to inductively represent prose texts in alternative ways, such as poetry (Keith & Endsley, 2020). They learn to extract poems from prose. Storytelling is a big part of hip-hop music; thus, blackout poetic transcription can be a relevant poetry teaching tool. Another version of this technique is when learners do not have to rely on the short story; instead, they can use their own short story as a basis of inspiration to write their own musical poems. A rubric can then be created by the educator to mark the poetry performance and written component (Creely, 2019).

4.3 Research Method

According to Denzin and Lincoln (2018), qualitative research aims to investigate phenomena in their natural setting to understand the kind of knowledge people bring to their work environment. Qualitative research was employed to explore the teaching of English poetry in a selected school in Gauteng, South Africa. According to Henning et al. (2005), qualitative research methods seek to determine the features, qualities, and properties of what is being explored. This justifies the need to triangulate the data gathered by examining multiple methods, such as document analysis, classroom observation, and interviews with learners and educators. According to Bhandari (2022), qualitative researchers collect and analyse non-numerical data, such as text, video, or audio, to understand concepts, opinions, or experiences. This implies that qualitative research is used to gather in-depth insights into problems or generate new ideas for research. Qualitative researchers are interested in understanding the meaning people construct to make sense of their world based on their experiences. Qualitative researchers come from different paradigmatic stances and apply distinctive and out-of-the-ordinary criteria for evaluating good research and the varieties of research contributions that are made. The qualitative method uses people's written or spoken words. Thus, using a qualitative research method, this study aims to describe and illuminate the participants' lived experiences. Cooper and Schindler (2014) succinctly note that qualitative research involves the researcher being immersed in the phenomena studied. The questions asked were basically to get an understanding of how the educators teach poetry; this was open-ended and there was no tight structure. With the learners, I mainly asked them how they were taught poetry and how they would like to be taught poetry. The learners' questions were more closed because they wrote down their answers on paper due to time constraints. The most relevant answers were chosen to help answer the research questions.

The qualitative research method was deemed necessary for this study as it afforded the following advantages (Dzvimbo, 2017): Flexibility: The data collection and analysis process can be adapted as new ideas or patterns emerge and are not rigidly decided beforehand. Natural settings: Data collection occurs in real-world contexts or naturalistic ways. Meaningful insights: Detailed descriptions of people's experiences, feelings, and perceptions can be used to design, test, or improve systems or products.

Generation of new ideas: Open-ended responses mean that researchers can uncover novel problems or opportunities that they would not have considered.

This chapter was guided by a normative study model. According to Routio (2007), normative models are used to describe existing problems and define improvements to the object of study. A normative research project proceeds through four successive stages: evaluation of the initial stage to understand the problem in context and define the need for improvements. This can be achieved through baseline surveys. Analysis of relationships and possibilities for change, wherein a literature study is conducted to provide guidelines. This could be a model formulated and intended for application to solve this problem. English poetry lessons were observed, and two educators were interviewed regarding their poetry teaching strategies.

4.3.1 Study Area

Winston High (pseudonym) is in a middle-class suburb in the Vaal area, Gauteng, South Africa. It is a former Model C school, thus, previously reserved for white English L1 learners prior to 1994. Since then, the demographic profile of the learners and staff has changed, as most white learners and staff left the school. Currently, the learner enrolment is predominantly black, as is the staff. In 2021 (when the research was conducted), 60% of the teachers were white and 40% were black. 99.9% of the learners were black, with only 0.1% being white learners.

This is a quintile 5 not-for-profit school; as such, it receives minimum assistance from the government. Although the learner population hails from different socio-economic backgrounds, most learners are from a middle-class background. The total number of learners in the school is 913, whereas the total number of teachers is 43. The learner-to-teacher ratio is 1:25. Each Grade has eight classes which at the time of the research adhered to COVID-19 regulations. The school uses English as the LoLT.

Most of the operational responsibilities at Winston High are taken care of by the school. These include the maintenance of the garden, sports fields, buildings, and provision of equipment for learners' games. The school is in a sound financial position, as it is able to hire extra teachers to keep the learner–teacher ratio low and improve academic success. The school has a library, which is not fully functional, with few reading materials. The school intends to convert this library into a computer laboratory to allow learners and teachers to access online resources. There is a pleasant staff room with a round table, couches, and a kitchenette. The school has an administrative team consisting of five staff members. The motto for the ethos of the school is 'Integrity'. Schools are considered to be well performing with more than 95% pass rates for their Grade 12 learners. Learners pay a monthly school fee of R850 (R8 500 per year).

Describing the background of the school is important because qualitative research aims to investigate phenomena in their natural setting to obtain a deep understanding of the kind of knowledge learners and teachers bring to the classroom (Denzin & Lincoln, 2018:3).

Two Grade 11 English HL teachers were purposively selected for this study. Data collection included observations and learner interviews on how they wanted to be taught. The lesson observations were recorded on video and transcribed, while the interviews with the educator were recorded on audio. Grade 11 learners answered questions on paper regarding their preferred poetry teaching methods. Grade 11 learners were purposively selected to participate in the study. The choice of Grade 11 was made because the Grade 11 CAPS English HL curriculum presents comprehensive poetry content (DBE, 2011b). Therefore, it is appropriate to investigate poetry teaching. The selected teachers were highly qualified. They both majored in English in their undergraduate degrees and have an honours degree as well as a professional teaching qualification. Both had 15 years of teaching experience in the same school.

The school was selected because its hidden curriculum is poetry, and the researcher has in-depth background knowledge of the school. Having prior knowledge as a researcher helps in accurately interpreting and analysing the data using qualitative methods and in using the normative study model to recommend methods for improving practice. Other schools in the area are mainly Afrikaans-based, so they were not suitable for answering the research questions.

The following actions were taken to ensure that ethical requirements were met in this study. Permission for the study was obtained from the Ethics Committee. Permission to conduct research in the selected school was obtained from the Gauteng Department of Basic Education (GDBE) and the school principal. Written consent to participate in the study was obtained from the two Grade 11 teachers and from the parents of the Grade 11 learners who were being observed. The learners provided their assent to participate in the study.

All participants were informed of the study's nature and purpose before commencement. Participants were assured of the confidentiality of data, and anonymity was ensured by the use of pseudonyms. Participants were informed of their right to withdraw from the study at any time or to decline to answer any questions without penalty. Acknowledging participants with their permission in all publications emanating from this study was assured.

4.4 Study Findings

4.4.1 Theme 1: Relevance of Current English Poetry Teaching Methods

Theme 1 explicitly showed three literary divides: how poetry is taught, how learners want poetry to be taught, and how poetry educators want to be perceived in terms of their teaching approach. The data revealed a misaligned inculcation of a socially constructive poetry curriculum for children. This is negative because it means that poetry education in schools is misrepresented for what poetry truly is, especially for the learners. The poetry curriculum further dims the light on learners' prior knowledge and skills they bring to school. The potential power of a socially constructive poetry pedagogic approach is discussed below.

Social Constructivism as an Effective Approach to Teach Poetry

Mr X (pseudonym), a teacher at Winston High, believes that he is an effective teacher when he links disciplinary knowledge in the school curriculum to the experiential knowledge of learners (Vygotsky, 1962).

Mrs Y (pseudonym) explained that she uses *"guided* 'enquiry' *instructions'."*

Mr X further elaborated.

> I choose certain songs, convert them into poems, and analyse them. I then used these songs as poems. This style, I think, assists learners to comprehend poetry given that they already know the songs…My learners come to understand how poetry is developed, and as such they are able to develop their own poetry. Remember songs are also poetry.

A socially constructed lesson does not just reflect the learners' mundane world; it adds value by assisting learners to apply the knowledge and resolve challenges in their local environments (Vygotsky's, 1962, 1978). Problems in learners' local lives may include poetry and songwriting skills.

Learner 1 explainedan opposing view to Mr X, she said that the poetry in class does not help her in anyway in her everyday life. She says that the poetry she listens to at home is "more interesting and some helps you to heal, helps to soften your soul.". Learner 2 says the "poems we learn at school are educational, but the ones I listen to at home in songs, talk about depression and anxiety."

The above is proof that, unlike Mr X's statement, the poems that learners are taught at Winston High are not socially constructive because they do not understand them and they do not improve their mental well-being on challenges like poetry understanding, poetry writing, anxiety, depression, or their soul that needs healing

(Learner 1 and 2). The data revealed that learners believe that the poems taught in school should improve their emotional intelligence and poetry understanding at the bare minimum.

Learner 3 says "If poetry was to made songs and we would taken time to listen to the songs, the poem would be much easier to understand." This statement does not corresponds to how Mr X explained he teaches poetry using transpoemation.

The hip-hop culture at Winston High is quite dominant amongst the learners, and it has been like that for decades. However, it seems that there are three literary divides: how poetry is taught, how learners want poetry to be taught, and how poetry educators want to be perceived in terms of their approach to poetry teaching.

Mrs Y's interview revealed that she does not have specific icebreakers in her poetry toolbox that help her reach lesson objectives; she, thus, appears to be teaching for assessment. The learners' feedback confirms that both Mr X and Mrs Y begin with the end in mind when teaching poetry to reach lesson objectives. Mr X wants to be perceived as a poetry educator who incorporates music into poetry teaching. In reality, according to the learners, he teaches to the test using traditional methods like Mrs. Y.

Mrs Y said

> I teach poetry to learners to interpret it accurately in preparation for the exams. Poetry is challenging for most learners, and as such, they tend to perform poorly in the examination, which affects their overall performance score in the English HL. I need to prepare them through tests and various exercises presented to them for better understanding.

Vygotsky (1978) explains that people's cultural backgrounds shape their ways of thinking and behaving. Thus, poetry at home seems to be more culturally educational than poetry taught at school because the prescribed poetry does not positively shape the learners' ways of thinking, behaving, and emotional intelligence, as it is not a reflection of their culture. As a result, learners' African cultural identities are not affirmed by the school. Learners are educated in school, yet at the same time, they are uneducated by the school; they are not empowered to be competent Hip Hop members of society. In this regard, Vygotsky's (1978) theory links culture and language. Mr X and Mrs Y's poetry teaching approach does not enhance learners' cultural identities where they see themselves in the school's poetry language; consequently, they are less likely to appreciate and respect themselves as Africans. This means that the hidden curriculum of Hip Hop that dominates Winston High is a passion that learners have, instead of educators, the HOD, and the Principal.

Theme 2:

This section outlines that learners at Winston High have multiple learning styles that are not accommodated by the school. Some are visual, others are audio-visual, and some are tactile, while a few prefer a text-based approach. The findings reveal that basing poetry lessons on text alone hinders universal literacy, which is a skill that learners should possess to navigate the real world.

Different Learning Styles in Poetry Education

Both Mrs. Y and Mr. X create the impression that mediation in poetry teaching is used to teach poetry successfully. The educators claim that different variables of mediation in poetry are usually used.

Mr X mentioned:

> The poems that we use in the classroom are not just text-based; we also have videos for each poem. I usually encourage learners to research the poem from Youtube, Sparknotes, music websites, and books before the lesson is taught."

Mrs Y expressed her honest views. She follows the following steps when teaching poetry.

> We read the poem and write important information on the board, and I then request learners to identify poetic devices. After that, we had a class discussion based on the poem.

Mr X claims to mediate poetry lessons in their classroom using English as the LoLT. Mr X says he prefers using YouTube videos and popular music, which is favoured by the youth, to teach poetry lessons.

Learner 4 explained that he would like poetry to be "taught in the same way, but with a touch of pictures." Learners 5 states "I want poetry to be taught in a way that is easier to understand e.g. making demonstrations." Learner 6 explains "I want poetry to be acted out by people, so that we can see people's emotions."

Mrs Y claims to use learners' prior knowledge of the subject to identify and correct misinformation. She further stated that she links previous lessons with the new lesson. I argue, based on evidence from learner interviews, that even with the linking, the learners are still not able to discern the thread that connects all her lessons. Learner 7 explains that hip-hop lyrics she listens to at home are "life lessons, I like many people to acknowledge." Learner 8 outlines that "at school they teach us poems of apartheid and "Shakespen" but in my spare time I listen to Hip Hop life lessons and love poems."

The findings indicate that both poetry teachers understood the importance of mediating poetry learning to enhance their learners' understanding. However, the data from the learners confirm that the teachers use traditional teaching strategies in class.

Learner 8 said this concerning inculcating hip-hop music in the class "no because we would lose focus." Learner 9 said that they want culturally relevant poems in the class, because "they would focus more." The above findings show that learners have different learning styles, and the majority of the audiovisual learners are not included in the poetry curriculum. Interestingly, a few who had a textile learning style were not interested in learning culturally relevant poems like hip-hop music. Learner 10 said, "the poems in school are different, because at school you are meant to get exposure."

Learner 11 said "The poetry I listen to at home is not educational and it is not appropriate for children." Learner 10 said "No, because some words and lyrics are inappropriate."

Recommendations: How Hip Hop can be Used to Improve Pedagogy In Teaching Poetry.

Approaching Poetry Teaching Through Hip Hop Pedagogy

It is recommended that the main source of poetry content should be the latest live poetry recitation of poetry that learners love and can relate to and that it should also be translated into text for learners who love written/traditional school poetry. Live poetry recitations should also accommodate learners who love to visualise and hear poetry lessons. Learners will develop a stronger emotional connection to poems. Live poetry reveals hidden messages that cannot be expressed in written text, such as body language, art, and emotions, which makes decoding poems easier for learners. This is difficult when teaching canons because original live demonstrations and culturally relevant videos of poets such as William Shakespeare are not available online to facilitate the learning. However, there are many "Hip Hop culture Shakespeares" that learners listen to at home that use the same literary devices that learners are currently tested on in exams. This shows that learners do meet the curriculum expectations on their own, but the teaching style and textbook do not create the required educational access to poetry knowledge.

Vygotsky (1978) emphasise the importance of moving from the known to the unknown to enhance understanding of the content. The nature of cultural symbols used in the poetry class is critical because cognition is activated and emerges through relevant socio-cultural interaction (Vygotsky, 1978). The way poetry is currently taught in schools is also unjust because it swims against the UNESCO objective of quality and inclusive education for all and the objective of the Constitution of South Africa, which is about the human right to basic education (Constitution of the Republic of South Africa, 1996b; UNESCO, 2025).

The Tragic Flaw of Hip Hop Pedagogy

Vygotsky (1978) explains that the zone of proximal development as the difference between what the learners can achieve on their own and what they can achieve with the help of a mentor. That is why street poetry should be included in the class, instead of letting learners listen to vulgar music on their own, so that the educator can scaffold and mediate their music-learning experiences. Additionally, to allow them to move through the zone of proximal development into higher-order thinking. Wertsch posits that learners go through social-level development before developing higher-order mental functions. This means that the educator should be there to lead learners into higher mental functions, stemming from the profane poetry in music. If the educator does not mediate the poems with profane language, someone else who is not a qualified poetry educator will miseducate them. In this way, the poems would be more of a learning experience, instead of a problem.

Emotional Intelligence

Composing hip-hop music in poetry is an important skill for learners to master, but what is more important is helping learners with depression, anxiety, and healing their souls (Learner 1 and Learner 2). When the learners' mental health is in check, their poetry writing skills in hip-hop music can be powerful. It is then that they will be able to read and interpret their own emotions, even if they are toxic, so that they resonate with and impact the society that consumes their lyrics. It is recommended that the school partner with an Educational Psychologist, especially an Educational Psychologist that use music therapy, so as to help them overcome the depression and anxiety the learners highlighted.

4.5 Conclusion

This article was initially political in nature. It was intended to be a type of activism, an advocacy for Hip Hop Pedagogy to be used to teach poetry. However, the researcher allowed the data to speak for itself. By triangulating sources of data, the article ended up being an outlet where learners cried for the poems in the classroom at Winston High to be more real. The article took its own direction, where learners are not asking for hip-hop music to be in their poetry classrooms; that would be a luxury. They just want the bare minimum: that their poetry lessons should be educational, emotionally uplifting, realistic, and based on pop culture music they love and relate to. The learners were also concerned that different learning styles were not accommodated in the poetry curriculum.

The above aligns with Sustainable Development 4 of the United Nations, which advocates for an 'increase in the number of youth and adults who have relevant skills, including technical and vocational skills, for employment and entrepreneurship' (UNESCO, 2025). The above goal can be easily achieved by increasing the relevance of poetry textbooks, teaching approaches, and musicalisation of poetry education in schools. At the moment, SG4 cannot be easily achieved because of the current English tradition, which limits the learners' lyrical development, "vocational skills, employment opportunities, and job creation" which is part of the 2030 objective of the United Nations (UNESCO, 2025).

References

Akinsola, M. K. (2014). Effects of game and poem enchanced instruction on pupils' achievement in mathematics. *International Journal of Education and Research, 2*(6), 373–386.

Akinkurolere, S., Adewumi, S., Aminu, S., Olowu, A., & Adetoro, A. (2021). *The implications of performance in the teaching of poetry at Rufus Giwa Polytechnic*, Owo, Nigeria.

Akom, A. A. (2009). Critical hip hop pedagogy as a form of liberatory praxis. *Equity & Excellence in Education, 42*(1), 52–66.

Biko, S., & Stubbs, A., I write what I like: A selection of his writings. (No Title).

Bhandari, P. (2022). *What is quantitative research? | Definition, uses & methods*. https://www.scribbr.com/methodology/quantitative-research/

Cooper, A. (2020). "That's Schoolified!" How curriculum, pedagogy and assessment shape the educational potential of poetry in subject English for black high school learners. *Education as Change, 24*(1), 1–20.

Cooper, D. R., & Schindler, P. S. (2014). *Business research methods* (12th ed.). McGraw-Hill Education.

Creely, E. (2019). 'Poetry is dying': Creating a (re) new (ed) pedagogical vision for teaching poetry. *The Australian Journal of Language and Literacy, 42*(2), 116–127.

Delchamps, V. (2018). Teaching poetry through dance. *Poetry and pedagogy across the lifespan: Disciplines, classrooms, contexts* (pp. 37–55).

Denzin, N. K., & Lincoln, Y. S. (Eds.). (2018). *The Sage handbook of qualitative research.* Sage.

Dewey, J. (1938). *Experience and education.* Macmillan.

Dewey, J. (2018). *Democracy and education by John Dewey: With a critical introduction by Patricia H. Hinchey.* Myers Education Press.

Donbavand, S. (2025). Can progressive education cultivate democratic dispositions?. In *Research handbook on education and democracy* (pp. 259–272). Edward Elgar Publishing.

Department of Basic Education (DBE). (2011a). *Curriculum and assessment policy statement (CAPS) grades 10–12 English first additional language (EFAL).* Department of Basic Education.

Department of Basic Education (DBE). (2011b). *Curriculum and assessment policy statement (CAPS) grades 10–12 English home language (HL).* Department of Basic Education.

Department of Basic Education (DBE). (2013b). Incremental implementation of African languages in South African Schools. *Draft Policy.* Department of Basic Education.

Dymoke, S., & McGuinn, N. (2021). Being allowed: Negotiating space for poetry writing with literature examination students. *New Writing, 18*(4), 440–451.

Dzvimbo, K. (2017). *Personal conversation with Kuzvinetsa Dzvimbo*, 3 March, 2017.

Emert, T. (2015). Pairing poetry and technology: Teaching from the "Outside Inward." *English Journal, 104*(4), 59–64.

Flores, T. T. (2021). Fighting to be heard: Latina adolescent girls writing toward change. *Journal of Adolescent & Adult Literacy, 65*(1), 65–73.

Freire, P. (1968). *Pedagogy of the oppressed.* (MB Ramos, Trans.) London.

Häkkinen, L. (2022). Runoja: Publishing site-specific poetry on a mobile application platform. *Master's degree.* Available at https://urn.fi/URN:NBN:fi:aalto-202206093680

Henning, E., Van Rensburg, W., & Smit, B. (2005). *Finding your way in qualitative research.* Van Schaik Publishers.

Keith, A., & Endsley, C. L. (2020). Knowledge of self: Possibilities for spoken word poetry, hip hop pedagogy, and "blackout poetic transcription" in critical qualitative research. *The International Journal of Critical Media Literacy, 2*(1), 56–82.

Lamb, C., Brown, D. G., & Clarke, C. L. (2017). A taxonomy of generative poetry techniques. *Journal of Mathematics and the Arts, 11*(3), 159–179.

Lambirth, A., Smith, S., & Steele, S. (2012). "Poetry is happening but I don't exactly know how": Literacy Subject Leaders' perceptions of poetry in their primary schools. *Literacy, 46*(2), 73–80.

Lawati, D. D., & Rana, K. (2022). Limbu poets' experiences of using facebook for promoting endangered indigenous language. *Journal of Underrepresented & Minority Progress, 6*(1).

Machado, E., Vaughan, A., Coppola, R., & Woodard, R. (2017). "Lived life through a colored lens": Culturally sustaining poetry in an urban literacy classroom. *Language Arts, 94*(6), 367–381.

Maduabuchi, C. H., & Maduagwu, U. M., Feasibility of using poetry in a textmasters' environment to teach mathematics to secondary school students.

Maduabuchi, C., & Maduagwu, M. (2015). Feasibility of using poetry in a texmasters' environment to teach mathematics to secondary school students. *International Journal of English Language Education, 3*(1), 104–114.

Margento, S. R., & Murat, T. (2021). *Various wanted. An (almost) missing original and five—Literary, computational, and visual—translations.*

Mavhiza, G., & Prozesky, M. (2020). Mapping pathways for an Indigenous poetry pedagogy: Performance, emergence and decolonisation. *Education as Change, 24*(1), 1–25.

McFarland, S. (2020). *Reading with the head and heart: A mindful analysis of poetry.* Curriculum unit, Charlotte Teachers Institute.

Mooney, B. (2022). To speak a true word: Remixing hip hop pedagogies, poetics, and literacies (Doctoral dissertation, Columbia University). Academic Commons. https://academiccommons.columbia.edu/doi/10.7916/df6j-ym97

Newfield, D., & D' abdon, R. (2015). Reconceptualising poetry as a multimodal genre. *Tesol Quarterly, 49*(3), 510–532.

Nicolau, F. (2022). Performativity of remixed poetry. Computational translations and Digital Humanities. *Open Linguistics, 8*(1), 401–413.

Okwor, I. N., & Chidiebere, G. R. (2021). Effects of transactional analysis method on senior secondary school students ability to appreciate poetry in Jos North, Plateau State, Nigeria. *KIU Journal of Social Sciences, 7*(2), 155–163.

Republic of South Africa (RSA) Constitution of the Republic of South Africa Act, no.108. (1996b).

Routio, P. (2007). *Collecting Normative Data.* Available at http://www2.uiah.fi/projekti/metodi/166.htm

Ugwuozor, F. O. (2020). Students' perception of corporate social responsibility: Analyzing the influence of gender, academic status, and exposure to business ethics education. *Business Ethics: A European Review, 29*(4), 737–747.

Ugwuozor, F. O., Ede, M. O., Abiogu, G. C., Ugwu, J. C., Ncheke, D. C., Nnamani, A., Albi-Oparaocha, F. C., & Obiweluozor, C. (2020). Effect of the constructivist method on junior secondary school students' achievement in poetry: implications for school counseling in Nigeria. *Global Journal of Health Science, 12*(5), 80.

UNESCO. (2025). *Goal 4: Quality Education.* Available at https://data.unicef.org/sdgs/goal-4-quality-education/

Vygotsky, L. (1962). *Thought and language.* Massachusetts Institute of Technology Press.

Vygotsky, L. S. (1978). *Mind in society: The development of higher psychological processes.* Harvard University Press.

Xerri, D. (2014a). Schools as "poetry-friendly places": Michael Rosen on poetry in the curriculum. *Arts Education Policy Review, 115*(4), 151–158.

Xerri, D. (2014b). The poetry of cities: On discovering poems in Istanbul, Sarajevo, and Bratislava. *Journeys, 15*(1), 90–108.

Part III
Student Success, Inclusion, and Support Systems

Chapter 5
Examining the Implementation of Extended Curriculum Programmes in a Rural-Based University in South Africa

Luyanda Marhaya, Munienge Mbodila, and Khashane Stephen Malatji

Abstract This study explores the implementation of Extended Curriculum Programmes (ECPs) at a rural university in South Africa. This study utilised a case study design and a qualitative approach. Data were collected through individual semi-structured interviews with lecturers and focus groups with students in faculties offering ECPs. The sample consisted of eight academics attached to faculties offering ECPs. In addition, eight students who had enrolled in ECPs were sampled, making the total sample size 16 participants. The findings revealed several themes related to the effectiveness of ECPs in enhancing academic support and fostering inclusive learning environments for students with disabilities. Participants highlighted the importance of tailored academic interventions and mentorship, which contribute to improved student engagement and retention rates, as well as overall socialisation. In addition, the findings indicate several challenges that affect the overall effectiveness of these programmes. Challenges, such as a lack of understanding of policies that inform the design of the programmes, a lack of awareness of funding mechanisms, and a lack of student development practices, were identified. This study underscores the need for strategic policy enhancements and collaborative efforts to optimise ECPs, aiming to improve academic outcomes and ultimately contribute to the realisation of quality education for students in rural contexts, as outlined in SDG4. Therefore, this study recommends an integrated student support programme that brings together ECPs with student development practices.

L. Marhaya (✉)
Vaal University of Technology, Vanderbijlpark, South Africa
e-mail: luyandam3@vut.ac.za

M. Mbodila
Walter Sisulu University, Mthatha, South Africa
e-mail: mmbodila@wsu.ac.za

K. S. Malatji
Faculty of Humanities, University of Limpopo, Polokwane, South Africa
e-mail: khashane.malatji@ump.ac.za

K. Dube and L. M. Tjabane (eds.), *Advancing Sustainable Development Goals in Higher Education*, https://doi.org/10.1007/978-3-032-27646-9_5

Keywords Extended programme provision · Higher education · Academic support · Inclusive learning

5.1 Introduction and Background

The Bantu Education Act of 1953 institutionalised segregation in education, resulting in the inferior provision of resources and opportunities for Black students compared to their White counterparts (Chiramba & Ndorifeni, 2023). Under apartheid in the early 1960s, in keeping with the vision of a separated society, the state created several colleges to cater to 'other' race groups, including the University College of the Western Cape for Coloureds and the University of Durban-Westville for Indians (CHE, 2024). Chiramba and Ndoferi (2023) assert that during the apartheid regime, communities were divided by race and geographical context to the extent that educational resources were unequally distributed, creating a huge rift between the white and black communities.

Discrimination in higher education is not only a South African phenomenon. For instance, In the United Kingdom, Lynam, Milani and Lafarge (2024), in their study, argue that HE environments in the UK remain 'white spaces' within which minoritised ethnic students face barriers that are not encountered by their White counterparts. Globally, the United Nations Education, Scientific and Cultural Organization (UNESCO) (1998:23) through its document, The World Declaration on Higher Education for the twenty-first century even proclaimed that admission to higher education should be based on the merit, capacity, efforts, perseverance and devotion, showed by those seeking access to it." Sit (2024) argues that the right to higher education is still not universal. Sit further argues that this is despite the inclusion of higher education in the United Nations' (2015) Sustainable Development Goals target 4, and its emphasis on ensuring equal access to affordable and quality higher education for all by 2030 highlights the importance of higher education for every individual around the world. UNESCO (2015) states that SGD 4 focuses on ensuring inclusive and equitable quality education and promoting lifelong learning opportunities for all. This means providing access to education at all levels, from early childhood to tertiary, and ensuring that the education provided is of high quality and relevant to the needs of individuals and society. The above goal underscores the importance of access to education at every level, particularly for this study, given the historic context of South Africa as a country.

In terms of redress in the South African higher education context, one of the first important pieces of legislation that recognised the imbalances during the apartheid regime was captured in the Education White Paper 3: A Programme for the Transformation of Higher Education. The Education Transformation culminated in a wide-ranging and extensive process of investigation and consultation that was initiated with the establishment of the National Commission on Higher Education (NCHE) in February 1995 by President Mandela, and the subsequent release of the Green Paper on Higher Education in December 1996 and the Draft White Paper on Higher

Education in April 1997 (DoE, 1997). The paper emphasised the need for equitable distribution of access and opportunity for students and staff along lines of race, gender, class and geography (DoE, 1997). When the new democratic government in South Africa came to power in 1994, there was a need to improve various sectors of society; among these sectors was the students' participation rate in higher education (Reddy, 2004). In 1994, 55% of students at public universities were black (African, coloured and Indian), 43% were African, and 55% were male; by 2010, these figures were 80% black, 67% African and 43% male (Bozalek, 2015). However, it has been realised that formal admission to higher-education programmes does not guarantee the success of students from historically disadvantaged backgrounds (CHE, 2024). Black students, as it was argued during this period, lacked the skills, conceptual background and language proficiency necessary to succeed at the tertiary level (Boughey, 2010). Students from educationally disadvantaged backgrounds have generally not been exposed to key academic approaches and experiences that are taken for granted in traditional higher education programmes (Scott et al., 2007).

In South Africa, for example, data in the form of the Council on Higher Education's (CHE's) Vital Stats series (CHE, 2023a, 2023b) show that, regardless of the university at which they are registered, the field of study, or the qualifications for which they are enrolled, Black South Africans fare less well than their White peers in terms of throughput and graduation rates (Boughey, 2024). This has resulted in several systematic interventions, as researchers such as Malatji et al. (2017) have pointed out various flaws in the widening participation gap in higher education. Consequently, Extended Curriculum Programmes with an Integrated Foundation Phase, now widely termed 'Extended Curriculum Programmes' ('ECPs'), were introduced in the early 2000s to promote access and success for students who had historically been denied access to quality higher education (Boughey, 2022). In this study, the terminology will be used interchangeably to refer to the same programme, as different periods of policy revision have vastly changed in terms of terminology. A key role of Foundation Provision is, therefore, to support educationally disadvantaged students who are underprepared despite meeting the minimum admission criteria by enabling them to be placed on an extended curriculum that will give them the academic foundations for completing their studies (DHET, 2012). Therefore, an Extended Curriculum Programme (ECP) is a formal undergraduate degree or diploma programme, the minimum duration of which is lengthened by half a year to a full year to incorporate substantial developmental academic provision additional to the coursework prescribed for the regular curriculum.

In the South African context, most extended programmes are located in centres for teaching and learning, similar to the global context, where they are regarded as the field of educational development, now often termed 'teaching and learning in higher education.' The naming of centres and units devoted to this endeavour has grown as institutions have looked to their practitioners to confront issues related to increased participation and the enhancement of student success (Boughey, 2024). Given the challenges articulated above, to achieve the aim of this study, the following research questions focus on policy implementation, the staff capacity to deliver these programmes, as well as student development practices at the institution, to ensure proper delivery of these programmes. Therefore, the objectives of this study are as follows, followed by the research questions that flow from the objectives:

Objectives of the Study

- To identify the policy-related factors affecting the implementation of the extended program provision.
- To assess the extent to which student development practices contribute to the satisfaction of students enrolled in extended programmes.
- To evaluate the capacity of academics to offer quality teaching and learning provisions to students enrolled in extended programmes.

Research Questions

The research questions necessary to support the primary research question are as follows:

- What policy-related factors affect the implementation of ECPs in rural universities?
- To what extent do student development practices contribute to the satisfaction of students enrolled in extended programmes?
- What is the capacity of academics to offer quality teaching and learning provision to students enrolled in extended programmes?

5.2 Literature Review

5.2.1 Policy Factors Affecting the Implementation of ECPs

Extended programmes were first recognised in state policy in 1997 in the Education White Paper 3: A Programme for the Transformation of Higher Education. Furthermore, the White Paper acknowledged the strategic role of extended programmes in fostering access to and success in higher education and committed the government to funding these programmes as 'integral elements of a higher education system committed to redress and to improving the quality of learning and teaching (Sect. 2.34) DHET, 2019 (Scott, 2009). Consequently, one of the first policy frameworks underpinning the development and funding of ECPs is the Foundation Provision in Ministerially Approved Programmes (DHET, 2012). Therefore, it is important to consider the policy's publication. The DHET (2012) guided institutions on the type of ECPs they could develop, which were:

- A 'fully foundational' course: this is a course that is preparatory to the regular first-level course in the subject(s).
- Extended course: combines regular course material with substantial foundational material and is substantially longer in duration than the corresponding regular course.

- An 'augmented' course: This covers all the material of a regular course and has the same duration but is taught separately and integrates substantial foundational material through additional formal contact time.
- Augmenting" course: This allows for the foundational elements of an augmented course to be provided as a separate (but integrally linked) course or module.

Furthermore, the Foundation Provision policy provides strategic guidelines on the structure, curriculum, and pedagogical approach of ECPs, ensuring that these programmes are designed to enhance students' academic skills and integrate them into mainstream university education.

Additionally, Motsoenyane (2020) highlights the critical importance of embedding academic support and skills development within standard coursework to ensure that students who do not meet traditional admission criteria are adequately prepared for higher education. Therefore, this integrated approach is particularly essential in rural universities, where students often encounter additional socioeconomic and educational barriers. Limited access to well-resourced schools, inadequate exposure to digital and academic literacy, and financial hardship contribute to the difficulties many rural students face upon entering university (Mukora, 2022). For these ECP programmes to be successful, the implementers must be aware of the policy prescripts that inform the role of ECPs in the broader South African context.

In addition, Boughey (2020) states that DHET does not micromanage programme design, although experts appointed to a Ministerial Reference Group scrutinise programme applications and may make recommendations for improvement. The design of ECPs is often left to the academics involved in them; if they are not aware of policies, it could prove detrimental to the success and sustainability of these programmes. The issue of funding mechanisms and policies for the regulation of ECPs is also important in discussing policy factors around these programmes. Furthermore, Boughey (2020) asserts that in 2004, the DoE published a set of criteria for the use of earmarked funding, and universities had to submit proposals for the use of this money, allocated to each institution based on the funding formula, in the form of 'Foundation Programme Grants. The funding mechanisms for foundation programmes are advanced in the Ministerial Statement for University Funding (DHET, 2023). The statement states that ECP students are funded annually in two ways. This is done by generating input subsidies within the teaching input sub-block grant (DHET, 2023) and distributing earmarked (ring-fenced) state funds for foundation purposes.

Bozalek (2015) states that the funds given to higher education institutions (HEIs) for ECPs were expected to 'improve success and graduation rates', particularly among disadvantaged students; however, the term disadvantage was not defined in the circulars informing universities about the funds. The real challenge, therefore, as posited above, is whether the policy frameworks created for the provision of ECPs cascade to the level whereby institutions can disburse the funds to an extent that it creates the transformation of a particular cohort of students as it was originally meant for in the various DHET policies, such as the Education White Paper 3: A Programme for the Transformation of Higher Education, Foundation Provision in Ministerially Approved Programmes, as well as the Ministerial Statement for University Funding.

There has been concerns over the implementation of certain policies within the higher education sector (Aruleba & Jere, 2022), in that there is poor dissemination of information about policy, limited awareness of policies, and a lack of awareness of the roles and responsibilities of implementation that flow from the policies. As a result, intended for this study, the extent to which the institution under study has developed ECPs that are compliant with the prescripts of the four models articulated in the DHET 2012 policy will be interesting. Conversely, academic understanding is fundamental to the design of these programmes so that they serve their intended purpose. Especially, given that ECPs play a vital role in bridging these gaps by providing structured academic support, foundational skills development, and an extended learning timeline to help students transition effectively into mainstream university programmes (Mkhize & Balfour, 2021; Motsoenyane, 2020).

5.2.2 *Staff Capacity is Geared Towards the Provision of ECPs*

Focusing on staff capacity for the provision of Extended Curriculum Programmes (ECPs) is essential for their success in higher education institutions. First, it is critical to ensure that academics possess the necessary academic credentials relevant to the subjects they teach. Therefore, the minimum qualification for academics teaching in ECP should be a Master's degree, depending on their level or academic rank. In addition, academics in this space require specialised training to reinforce their pedagogical skills, learning support strategies and inclusive teaching practices. Regular workshops and seminars are also critical for updating staff on best practices and new teaching methodologies. CHE (2020) states that most developmental provisions in ECPs are taught by Academic Development (AD) practitioners, and the links between AD practitioners responsible for teaching developmental courses and subject specialists employed in mainstream departments are often tenuous. This does not bode well with the intent of ECPs, in that it serves as a preparatory introduction so that students who are enrolled in it should not struggle with adaptation when they begin to take the regular first-year curriculum. Additionally, Ogude et al. (2021) found that academic staff in ECPs did not have a forum to discuss student performance in all modules or per academic stream compared to their mainstream counterparts. This points to significant differences between the value of ECPs and the operations in mainstream qualifications.

Garraway and Bozalek (2019a, 2019b) assert that general university subsidy, rather than in three-year cycles, will go some way to creating more permanent FP posts, and thus enable more systematic academic staff development. Bozalek (2015) found that in one disadvantaged rural university, the greatest struggle was to attract academics to teach on these programmes, which were seen as having lesser value. Systemic challenges, as alluded to above, seem to be the most severe in the implementation of ECP provision. However, Shay and Wolff (2016) pointed to various systematic imperatives that should be in place in the implementation of ECPs. The first is leadership and ownership, where they found that the most successful ECPs were those in which all heads of departments (and, to some degree, deans) strongly supported the ECPs. The second is organizational arrangements, where they found that the more successful extended programs appeared to have greater organizational permeability between mainstream academics and AD staff (Shay et al., 2016).

5.2.3 Student Experiences on Development Practices in the Institution

Creating the best possible student experience in Extended Curriculum Programmes (ECPs) involves several critical elements based on research. First, individualised learning support is important for tailoring academic assistance to meet diverse learning needs. Moreover, the use of active teaching strategies that create space for critical engagement encourages students to prepare themselves before class (flipped classroom). Moreover, the curriculum must be delivered in a way that assists students in relating what they are learning to real-life situations. Moreover, creating an inclusive environment for students is important to ensure a welcoming environment that values diversity and fosters a sense of belonging (Kgwete & Malatji, 2021).

Students enrolled in ECPs often encounter a range of academic, social and psychological challenges that can hinder their progress and overall success (Mthethwa, 2022). Many of these students come from under-resourced backgrounds and face difficulties in adjusting to the demands of higher education (Baleni et al., 2016). These challenges include inadequate academic preparedness, financial constraints, limited digital literacy and difficulties in accessing essential learning resources. Given these obstacles, institutions must adopt tailored interventions that holistically support ECP students, ensuring that they receive the necessary academic, emotional and institutional support to thrive (Mokganya & Zitha, 2024).

Mampane (2021) highlights the critical need for customised management approaches that cater to the unique needs of ECP students. This study discusses the development of the Mamelodi Referral System, a structured intervention aimed at providing targeted academic and psychosocial support. This system facilitates the early identification of at-risk students, allowing institutions to implement proactive measures, such as mentorship, additional tutoring and student engagement initiatives (Cassells, 2018). This study underscores the importance of adopting a student-centred approach to managing ECPs to enhance retention and success rates. Furthermore, mental health has emerged as a significant factor influencing student performance in the ECPs.

Mutinta (2022) examined the psychological well-being of ECP students at a South African university, revealing that many face heightened levels of stress, anxiety and depression. These mental health challenges often stem from academic pressures, socioeconomic difficulties and the struggle to integrate into university culture

(Begum et al., 2024). This research emphasises the urgent need for comprehensive support systems that extend beyond academic assistance to include mental health services, peer support networks and wellness programmes (Shalaby & Agyapong, 2020). Therefore, institutions implementing ECPs must integrate mental health awareness initiatives and counselling services into their student support frameworks to promote holistic well-being and academic success. Addressing these challenges requires a multifaceted approach that combines academic development with psychological and institutional support (Christiansen & Slammert, 2006). Universities must invest in mentorship programs, student wellness centres, and digital learning resources tailored to the needs of ECP students. This fosters a supportive and inclusive academic environment, and institutions can enhance the overall effectiveness of Extended Curriculum Programmes, ensuring that students are not only academically competent but also mentally resilient and socially integrated into the university community (Aithal & Aithal, 2023).

Evaluating the effectiveness of ECPs is crucial to ensure that these interventions achieve their intended objectives of improving student success, retention and graduation rates (Johnson, 2017). Given that ECPs are designed to support students who may not have met the traditional entry requirements for higher education, continuous assessment and refinement of these programmes are necessary to maximise their impact. Institutions must implement rigorous evaluation mechanisms that consider student performance, engagement levels and overall academic development to determine the strengths and weaknesses of the ECPs. Research has shown that while ECPs have contributed to improved retention and progression rates, there remains a continuous need for assessment and enhancement to address the evolving student needs. Scott (2020) highlights that extended programmes not only support individual students but also serve as a broader framework for understanding and addressing challenges within South Africa's higher education system. These programmes provide valuable insights into student learning patterns, institutional support structures, and the effectiveness of teaching and learning strategies (Vermunt & Donche, 2017).

By analysing data from ECPs, universities can refine their pedagogical approaches, enhance student support services and inform national policies aimed at improving higher education outcomes (Johnson, 2017). Johnson further suggests that the success of ECPs is influenced by various factors, including curriculum design, teaching methodologies, student support mechanisms and institutional commitment to inclusivity. A well-structured ECP should integrate subject-specific foundational courses with academic literacy, numeracy and personal development components to ensure that students receive holistic support. Moreover, the role of lecturers and academic mentors is critical in fostering an engaging and supportive learning environment that encourages persistence and motivation. To ensure the continued effectiveness of ECPs, universities must implement data-driven evaluation strategies that track student progress over time (Emma & Owen, 2025). This can include monitoring pass rates, dropout rates, time-to-degree completion and qualitative feedback from students and faculty members. Additionally, collaboration between academic departments, student support services, and institutional researchers is essential for identifying best practices and areas for improvement. Eventually, the lessons learned from ECPs extend beyond individual students and institutions, offering system-wide insights that can inform national education policies and strategies (Ehren & Baxter, 2020). By leveraging research and evidence-based improvements, South African universities can strengthen the role of ECPs in promoting equitable access to quality higher education and address the broader structural challenges within the sector.

For ECPs, tracking systems serve a specialised function (Parés et al., 2024). These programmes often cater to students who do not meet regular entry requirements and may need additional support to bridge the academic gap between their prior knowledge and university-level demands. The tracking systems used in ECPs are designed to monitor not only academic performance but also engagement with supplementary academic resources, workshops and support services specifically tailored to students' needs (Mampane, 2021). Tracking systems in ECPs are vital for identifying early warning signs of student struggles, such as poor performance in foundational courses or a lack of participation in remedial workshops. This early identification enables universities to provide customised interventions, such as academic tutoring, mentorship programmes and peer support initiatives. The systems also track students' attendance in supplementary courses that focus on foundational skills such as academic writing, study skills and numeracy. This allows educators to assess whether these interventions are effective in helping students build the skills they need to succeed in mainstream courses (Mampane & Khoza, 2020). Moreover, tracking systems can support the development of personalised learning pathways for students to follow. For example, if the system identifies a student who is struggling with specific academic skills, an advisor can work with the student to create a customised learning plan that includes targeted support such as additional tutoring, time management workshops, or specific reading materials (Ngwenya & Ramphal, 2023a, 2023b). The integration of tracking systems into ECPs also enhances communication between academic staff and student support services, creating a more holistic approach to student development.

5.2.4 Theoretical Framework

5.2.4.1 Kolb Learning Styles in the Context of Extended Curriculum Programmes

This study adopted Kolb's learning theory to unpack the extent of pedagogical practices within extended programmes (Kolb, 1974). Each theory presents a different learning style that lecturers should consider for the total inclusion of diverse students in ECPs. Kolb's theory lenses present different learning attributes that need to be considered during the transition from basic education to higher education. In addition to the academic jargon that lecturers need to introduce to students, they also need to consider their different abilities to complement their strengths and weaknesses for the preparation of the academic programme.

The main purpose of extended programme provision is to bridge the student articulation gap and prepare students for academic programmes. Therefore, to prepare students for their academic journeys, lecturers within ECPs need to consider different ways in which students learn to unlock their full potential. According to Malatji et al. (2023), several factors influence students' preference for a particular learning style. Therefore, in this study, the researchers looked at how different Kolb's theoretical lenses are connected and how they can be integrated with extended programme provision.

Kolbs (1974) articulates how pedagogical practices within ECPs should be carried out to accommodate students with different learning abilities to prepare them for the formal academic programmes. Therefore, this theory advocates that in ECPs, lecturers need to use multiple ways of engaging students and support them from different angles. For example, students who use converging learning styles may prefer a different teaching strategy compared to students with diverging learning styles. Hence, in this study, the researchers looked at policy-related matters and how they affect the implementation of ECPs, the extent to which student development practices contribute to the satisfaction of students enrolled in extended programmes, and the capacity of academics who offer quality teaching and learning provision to students enrolled in extended programmes. Therefore, to achieve the objectives of this study, Kolb's learning theory is useful as it prepares lecturers in ECPs to understand how students learn for a smooth transition from high schools to higher education institutions.

5.2.4.2 Application of Kolb's Theory in the Context of ECPs

Diverging (Feeling and Watching—CE/RO)

Students with diverging learning styles are sensitive to their surroundings. Lecturers in ECPs need to accommodate such students in their pedagogical approaches. According to Kolb (1974), students with this learning style learn through observation or watching things, as opposed to doing them. In this study, ECPs must be structured to accommodate students with this learning style. During the process of watching, students collect information that assists them in building their understanding of the topic. According to Malatji and Singh (2018:98), 'learners in diverging learning style perform better in situations that require idea generation, for example, brainstorming'. In the context of this study, lecturers need to use teaching strategies that create space for students to brainstorm through audio-visual resources to allow them to attach meaning and generate understanding. When this learning style is accommodated in ECPs, it increases the chances of students succeeding in their academic programmes.

5.2.4.3 Assimilating (Watching and Thinking—AC/RO)

The second theoretical lens, as articulated by Kolb (1974), is assimilation. Students with an assimilation learning style prefer to learn content in a logical manner. To accommodate students with this learning style in ECPs, lecturers need to follow Bloom's taxonomy to arrange the content logically from simple to complex. According to Malatji et al. (2023), students with this learning style are good at understanding the widespread of information and synchronising it clearly and logically. In the context of this study, once the students' articulation gap is addressed, it results in a smooth transition from high school to university. Therefore, ECPs students need to be gradually introduced to the academic space, including academic writing and ethical issues within it.

5.2.4.4 Converging (Doing and Thinking—AC/AE)

The next lens is converging. In this learning style, converging learners prefer doing and thinking in their approach to learning. Students with a convergent learning style are strong problem solvers. This learning style is critical for ECPs to prepare students as critical thinkers and independent students. Kolb (1974) argues that students with converging learning styles can use their problem-solving skills to solve practical problems in their surroundings. In this study, once students master problem-solving skills, they will be good at communication, which addresses their articulation gaps. Critical thinkers are likely to succeed in their academic programmes. Malatji and Singh (2018:99) argue that students with a converging learning style 'prefer technical tasks and are less concerned with people and interpersonal aspects'. In ECPs, students with a converging learning style are best at finding practical uses for ideas and theories. In other words, they can solve problems and make decisions by finding solutions to questions and problems that arise.

5.2.4.5 Accommodating (Doing and Feeling—CE/AE)

The last theoretical lens, as presented by Kolb (1974), is accommodation. Students with accommodating learning styles prefer doing and feeling things. In this learning style, students prefer hands-on learning as opposed to theoretical learning. In the context of ECPs, students may be exposed to hands-on practical sessions as opposed to a series of information. In any curriculum, some topics require students to engage in practical activities, and ECPs should prepare students for such activities. Practical activities need to be coordinated in a very cooperative manner that will assist students in engaging with one another and improving their articulation gaps. Students in this learning style are 'attracted to new challenges and experiences, and to carrying out plans' (Malatji & Singh, 2018:99). This learning style requires students to have critical thinking and analysis skills. Therefore, this learning style is very relevant for ECPs and good for preparing students for official academic programmes. The following section discusses the methodological issues of this study.

5.3 Methodology

This study has followed a qualitative approach to explore the implementation of ECPs at a rural-based university in South Africa, focusing on its implications for student academic development practices, policy imperatives, as well as staff capacity (Megbowon et al., 2023). The interpretive paradigm was used to explore the day-to-day experiences of lecturers and students in ECPs (Marais, 2022). Utilising a case study design and qualitative approach, data were collected through individual semi-structured interviews and focus groups involving lecturers and students in faculties offering extended programmes. The use of a case study as a research design enabled the researchers to examine the university in a rural context to explore the complexities of implementing ECP in a rural setting. In addition, the exploratory case study revealed specific challenges associated with the implementation of ECP with a view to suggesting future initiatives. Through a case study design, the researchers were able to assess the ECP and its effectiveness in student learning and retention in rural universities. Purposive sampling was used to select 16 participants, who were drawn from eight academics (two Associate Professors, four Senior Lecturers, and two lecturers) attached to the faculties of Science and Engineering and the Built Environment that were offering extended programmes, and eight students who had undergone an extended programme.

Researchers found the sample of 16 to be more manageable to allow critical analysis of the emerging themes through thematic analysis. The sample size of 16 allowed the researchers to reach the saturation phase, which led to a critical analysis of the findings with a view to suggesting actionable recommendations. Participants were drawn from less experienced (lecturers with no PhDs) to more experienced ones, which consisted of Senior Lecturers as well as Associate Professors. The aim was to bring a balance of responses to establish whether experience impacts the success of ECPs. Individual semi-structured interviews were conducted with lecturers, while focus group interviews were conducted with students. Both data collection instruments were piloted, and necessary adjustments were made to improve them. Participation in this study was voluntary. Both lecturers and students were given consent forms to complete, and they had the right to withdraw from the study at any given time. All information provided was treated as confidential, and the researchers ensured the anonymity of the participants.

Data were analysed using thematic analysis. Regarding ethical considerations, permission to conduct the study at the university was granted, and participation in the study was voluntary.

5.4 Results of the Study

The results presented in this study are based on the following themes: Awareness of national policies and their reflection on how they are used in ECPs; funding policies that inform the funding of students enrolled in ECPs; staff capacity of lecturers involved in the teaching of ECPs; staff development initiatives geared towards ECP academics; alignment of their teaching with the mainstream curriculum and staff; management of ECP teaching and faculty leadership; student development practices affecting the implementation of ECPs; and socialisation into the university culture.

The question on policy factors affecting the implementation of ECPs was targeted at the eight academics who were involved in the teaching and learning domain. While the researchers could have opted to pose questions to the university policymakers, the rationale behind isolating the academics was the assertion that policy is everyone is business, as it informs their knowledge of ECPs from a policy level, so that they are aware of the dynamics of policy implementation as well as how it ultimately affects their teaching and learning. These were organised through semi-structured interviews. The questions posed to students also aligned with the theoretical framework advanced by Kolb (1974) in terms of the following themes: diverging (feeling and watching—CE/RO); assimilating (watching and thinking—AC/RO)'; converging (doing and thinking—AC/AE) and accommodating (doing and feeling—CE/AE).

The results show that academics are not knowledgeable about the policy frameworks underpinning ECPs at different levels.

5.4.1 Awareness of National Policies and Their Reflection on How They Use These in ECPs

Participant One Opined:

> *I was never really oriented to the national policy frameworks, no one in the department told me of any policies, other than that these students were not prepared to enter at the first-year level, and therefore I just needed to teach them foundational skills.*

Participant Two Indicated:

> *I have heard of the White Paper about the transformation of students enrolled in ECPs, but exactly how it relates to my day-to-day teaching, I was not clear, as no one bothered to take me through.*

Three participants were observed.

> *Since my studies were in Higher Education Studies, yes, I have read some policies, but not about the ECP that I teach.*

Participant Four:

I have no idea what these policies are and how they relate to how I teach, as I am responsible for the content that I teach.

Participant Five:

I know some of the policies that you are talking about, but there is no institutional policy that is in existence that I was told about.

All the quotes above show the difference between what the national policies state as the underpinning importance of being involved in these programmes. The lack of awareness of the salient aspects of policies that emphasise the different models of ECPs. In addition, the crucial aspects of how these policies in some way or the other influence how they carry themselves in knowing the value of an ECP at a macro level is a concern to all stakeholders involved i.e. the broader institutional imperatives, the students they teach wherein they have to drive the agency of ECPs at a level of teaching but also for themselves as academics to be well versed so that they at a agents of this transformation that is supposed to inform their delivery in teaching students enrolled in ECPs. If academics who teach in ECPs are not aware of policy imperatives, there would be obvious gaps in their understanding of policy and their teaching and learning terrain.

5.4.2 Funding Policies that Inform the Funding of Students Enrolled in ECPs

Funding policies towards the implementation of ECPs is also a significant area that academics involved in these programmes must be aware of so that they are conscientised to the critical value and their understanding of their roles, but most importantly, how they fit into sustainability as academics who must be retained, trained, and developed so that the provision of the programmes is a success.

Participant One:

I know that there is a three-year cycle in these programmes because we signed three-year contracts, and we are always told that our contracts will be renewed as and when the department (referring to DHET) provides funding for the next cycle.

Participant Two:

I feel that because of the funding nature of these programmes, which is never explained to us, is problematic; we are never made to feel like we are part of the university.

Participant Four:

I have no idea how these programmes are funded, as a result, I am always looking for a lecturer position in the mainstream programme.

The disjuncture illustrated above indicates that the lack of awareness of funding mechanisms for the academics who teach in these programmes shows that often, these academics are thrown into these programmes with the idea that eventually they will be part of the mainstream structured and funded programmes, which can secure their jobs at the university. The idea that they must sustain the knowledge gained from their experiences in the tenure that they are involved in ECPs is not their concern as the focus is monetary and sustaining their academic careers rather than ensuring that these programmes mature and are sustained in such a way that they are successful, not only for the university but for the students as well.

5.4.3 Staff Capacity in Lecturers Involved in the Teaching of ECPs

Staff capacity and development in ECPs is a crucial issue as they deal with students who must transition to mainstream qualifications with better adjustment skills than those they may have inherited from their secondary school level. This was also targeted at the academics who teach in the ECPs, with the intention of ascertaining their preparedness to teach in the ECPs (Kgwete & Malatji, 2021). This was done through semi-structured interviews.

5.4.4 Staff Development Initiatives Geared Towards ECP Academics

The professional development of academic staff is essential, especially in the constantly changing environment of higher education and Extended Curriculum Programmes (ECPs). It is crucial to provide staff development programmes specifically designed for ECP academics to equip lecturers with the skills and knowledge they need to properly support a wide range of students. These programmes help bridge the gap between potential and achievement in higher education by promoting a culture of ongoing learning and cooperation, which in turn improves teaching methods and student outcomes. This emphasis on staff development highlights the dedication to creating an inclusive and responsive academic community, which is necessary for the effectiveness of ECPs in addressing the specific challenges faced by students from diverse backgrounds.

Participant One Stated:

> *Yes, because we are part of the faculty and indeed all staff development initiatives are publicised to all staff, we do attend the workshops on issues around curriculum, assessment, teaching and learning strategies, but these remain generic to all staff in the faculty.*

Participant Four:

> *I am aware of the staff development initiatives, and I do attend them as and when they are available.*

Participant Eight:

I am aware, our HOD emphasises that we need to attend them, I guess it is up to one to dedicate time for these initiatives.

ECP academics must have staff development programmes to improve their teaching effectiveness and meet the various needs of their students. These programmes encourage a culture of teamwork and responsiveness by fostering continuous education. At the end of the day, investing in staff development results in better student outcomes and more success in Extended Curriculum Programmes.

5.4.5 Alignment of Their Teaching with the Mainstream Curriculum and Staff

Participant One Opined:

Since we develop our module outlines and learner guides, we rarely have meetings with our mainstream counterparts to talk about gaps that we need to fill in developing our ECP curriculum.

Participant Four:

No, I have never sat with any mainstream lecturer to discuss the modules that are related to the one that I teach, other than hearing issues that concern us at the faculty board level.

Participant Six:

I have met the lecturer involved in mainstream only at departmental meetings of faculty meetings; we have never really sat down to discuss the content of our modules and how they relate.

This remark emphasises a major disconnect between instructors of mainstream and Extended Curriculum Programmes in terms of cooperation and communication. Missed opportunities to integrate viewpoints that might improve the curriculum are indicated by the paucity of thorough discussions of module content. More dialogue and interaction are necessary to guarantee alignment and coherence in instructional strategies, which will ultimately improve student learning outcomes.

5.4.6 Management of ECP Teaching and Faculty Leadership

Managing an ECP programme differs from managing a mainstream qualification. Mainstream programmes are already established, form part of the broader Programme Qualification Mix that is approved by the DHET, accredited by the CHE, and registered by the SAQA. ECPs take a different route, and as shown in the literature in this paper, their recruitment is crucial to the sustainability of an ECP programme, but also to faculty-wide management.

Participant One:

> *Since we are part of the faculty, we have our own HOD who normally cascades issues discussed at that level to us. So, to some degree, I feel we are represented.*

Participant Four:

> *Yes, indeed, our HOD is present, although sometimes I feel that we are a bit isolated from the broader faculty. Sometimes I do feel a sense of isolation, as if we are still teaching in secondary school, I cannot explain it to you in words (the interviewer did probe on this, but the participant was not coming through with the exact meaning).*

Participant Six:

> *We are part of the faculty board, but we normally just sit there, and we do not raise our issues, and it is normally up to the HOD to raise our frustrations at that level. He does represent, but because he is also involved with his colleagues at the mainstream level, maybe he downplays our issues.*

5.4.7 Student Development Practices Affecting the Implementation of ECPs

Student development practices are fundamental to the success of ECP students in higher education. These programmes seek to assist students in adjusting to a university life that is unknown to them, especially the deserving students who are selected for ECPs. These were focus group discussions with senior students who had undergone the ECP programme. The intention of selecting the participants was not necessarily based on whether they had been exposed to any form of preparatory student development practices, but rather on general practices which must be in place to assist them outside their normal classes. The extent to which any student development practices in place and those not in place at the university under study had any value or whether the participants simply did not know them was also considered in the study.

Since these were focused on group questions, students were grouped into the S category (representing divergent students, 1–8).

Broadly, the focus group questions centred on the following theme:

- **Academic Student Development Practice Interventions:**

The questions in this section were as broad as possible, covering the different main aspects discussed in the literature review section on students' development practices.

5.4.8 Student Development Practices (Mentorship, Tutorship)

Mentorship and tutoring are examples of student development methods that are essential for creating a nurturing learning atmosphere that promotes academic and personal development. In addition to providing students with direction and support, these methods foster meaningful connections that may have a profound impact on their learning outcomes. By prioritising mentorship and tutorship, colleges may help students overcome obstacles, acquire vital skills and realise their full potential, all of which will ultimately contribute to their overall success in higher-education. The participants in this group responded as follows:

S 4 Stated:

> *I think the thing that helped me was the fact that senior students would often tell us how they had dealt with some of the stuff*,

S 3 Shared the Same Sentiment:

> *I think to add on that, we got to know how they approached difficult stuff in their modules.*

S32 Said: "*And it was easy to ask them in class if we were struggling*"

S 1 Stated:

> *I am glad I went through the programmes; it gave me foundations in some of the modules that I am even doing now in the second year.*

S 2 Said:

> *To add on that, I did not do Accounting in high school, but the foundations that I learnt from have meant that I can adjust well.*

S 4 Stated:

> *I think that it has assisted me with the foundations of the courses that I am currently doing.*

S 5 Said:

> *tutors gave us exercises that helped us to prepare for exams; because of this, I was always prepared for my exams.*

Tracking student performance to have targeted interventions for those who may be categorised as at-risk students is an integral part of the family of student development programmes. This new phenomenon in higher education has the potential to positively affect students' performance. However, the respondents indicated that they were unaware of any such intervention at the university. Given this background, it was clear that there seems to be a policy gap in some of the initiatives that were supposed to be embedded for the successful elimination of ECPS, as observed from the interviews below:

S 1 Said:

No, there isn't. There is no tracking whereby you enrol in the foundation of mainstream 1st year, and then there's no tracking, there's no one who keeps track of your doings or how you are performing, besides you and your family and those who are just asking you, but not for a programme that is running in the school, there's none.

S 3 Stated:

There's no such thing, there has never been, but I think the University should consider it, since this happens in other universities.

5.4.8.1 Socialisation into the University Culture

S 3 Stated:

As for the socialising part, there's not much of socialising, but we are more like given a tour of the school, like we are only given a tour, for example, this is our finance building, this is the School of Meds, this is the library. Because most of the time we were given a tour of the library, that's the most common place that people are taken to, the library, and shown how the system of the library works.

S 4 Said:

Most of the orientation, whereby socialising with the staff, it's not there. You get shown that this is the School of Mathematics, but I would not know the departments inside or what, yes.

S 8 Stated:

I am now better organised than I was as I used better time management that I learnt in my first year of foundation.

S 7 Concurred That:

I know now how to skim text, summarise and create an argument, but most of all, I also developed self-awareness skills that have assisted me even as a person.

S 6 Stated:

I think I am a better person now; I do not know how I would have adjusted to university if I had not done these courses; they helped me so much.

It became clear from the students' sentiments that they had gained a lot from these soft skills modules. The other key aspect that was picked up was that the modules assisted them with adjusting, which could bode well for them being retained in the university and for them transitioning to other senior years. Tracking systems are particularly important at the entry level to ensure that appropriate interventions are made in time (Balfanz & Byrnes, 2019). This does not exist in the institution.

5.5 Discussion of Findings

The findings reveal that there is no awareness of national policies that inform the conceptualisation of ECPs, which also implies that academics teaching in ECPs are left to decide what is taught in these programmes without any thought on the design of these programmes. Therefore, the entire process is determined by teachers' experience which Kolb (1974) refers to as concrete experience, and the decision of how to teach relies on lecturers' feelings. These findings align with the CHE (2022) assertion that the implementation of certain policies within the higher education sector is poor in terms of disseminating information about policy, limited awareness of policies and a lack of awareness of the roles and responsibilities of implementation that flow from the policies. This lack of awareness is also worrisome, as ECPs must be conceptualised in such a way that they are responsive to the different needs of the students enrolled in them. If there is no lack of awareness, this could imply that academics have the free will to develop non-responsive ECPs. The DHET (2012) policy specified the type of ECP institutions that could be developed. Without this awareness and a lack of a targeted and responsive ECP policy, whatever is taught at the ECP level might not be as responsive to the key tenets of the policies, which underscore the transformative nature of these programmes.

ECPs do not necessarily follow the sequence of formal programmes, where these programmes must be accredited by the CHE and SAQA registered, as they are approved by the DHET. However, with that approach, as indicated by Boughey (2020), the DHET does not micromanage programme design, although experts appointed to a Ministerial Reference Group scrutinise programme applications and may make recommendations for improvement. Academics who teach in terms of awareness of ECPs programmes also flagged a lack of awareness of the funding of ECP (Mgutshini, 2022). This means that academics might not be well-versed with the entire funding formula used by DHET, and from the interviews, emanating from this lack of knowledge, most academics indicated that their aspirations are targeted towards their career sustainability. This confirms Bozalek's (2015) finding that the funds given to higher education institutions (HEIs) for ECPs were expected to 'improve success and graduation rates, particularly amongst disadvantaged students. However, the term disadvantage was not defined in the circulars informing universities about the funds (Bozalek, 2015). If academics are not aware of the funding imperative of ECPs, the intention of the DHET in introducing ECPs, and the lack of their understanding of assisting 'disadvantaged students'. This implies that they might be disconnected from the notion of what these programmes are supposed to achieve.

Staff capacity interviews seemed to yield positive results regarding the involvement of ECP academics in the management of ECPs, as they are well represented in faculty decisions. This is consistent with Shay et al.'s finding that the most successful ECPs were those where all heads of departments (and to some degree deans) strongly supported the ECPs. Secondly, organisational arrangements, where they found that the more successful extended programmes appeared to have greater organizational permeability between mainstream academics and AD staff. Kolb (1974) argues that converging lens allows lecturers to think and brainstorm together before the implementation. ECPs differ in their approach to historic design and transformation imperatives. There seems to be no relationship between the mainstream and ECP curricula. This is confirmed in Ogude et al. (2021), who, in their findings from their study, found that academic staff in ECPs did not have a forum to discuss student performance in all modules or per academic stream compared to their mainstream counterparts.

Student experiences had differing views, where they pointed out that the different programmes did assist them in preparing them for the mainstream curriculum (Kolb, 1974). This is confirmed by Mampane (2021), who highlighted the critical need for customised management approaches that cater to the unique needs of ECP students. However, there seems to be no monitoring of these programmes through a targeted tracking system that can identify loopholes before students struggle with their studies. This is emphasised by Mampane (2021), who states that the tracking systems used in ECPs are designed to monitor not only academic performance but also engagement with supplementary academic resources, workshops and support services specifically tailored to these students' needs.

5.6 Recommendations

The authors recommend that a broader policy discussion on the importance of these programmes is needed. This not only ensures that academics involved in ECPs understand the background of these programmes, but also assists them in understanding the teaching and learning and pedagogical value of being involved in these programmes, as proposed by a 2023 study on ECP (Mouton, 2023).

Institutions must also be transparent about the funding mechanisms of these programmes so that they are better equipped to understand the nuances of national-level funding. This will go a long way in ensuring that these programmes at the institution are sustainable and can mature in terms of delivery and curriculum relevance. Student development practices must be tailored to track these students; therefore, a proper system must be implemented at the institution.

5.7 Conclusion

The study concludes that a lack of policy awareness among academics who teach in ECPs can severely impact the proper development of ECPs, similar to Ayabei (2024). Additionally, the lack of funding mechanisms that must permeate to the level of academics who teach in ECPs can have severe implications for the sustainability of these programmes. Staff capacity is also commendable, but the study concludes that there seem to be no targeted interventions that solely target ECP academics. We also conclude that there are appropriate student development practices that ensure that students enrolled in these programmes adjust to the university culture, albeit the other missing practices such as tracking and monitoring mechanisms that must be in place to ensure that these students are well monitored and tracked to ensure seamless transition to mainstream programmes.

References

Aithal, P. S., & Aithal, S. (2023). How to increase the emotional infrastructure of higher education institutions. *International Journal of Management, Technology, and Social Sciences, 8*(3), 356–394.

Aruleba, K., & Jere, N. (2022). Exploring digital transforming challenges in rural areas of South Africa through a systematic review of empirical studies. *Scientific African, 16*, Article e01190.

Ayabei, C. J. (2024). Exploring Factors Influencing Absenteeism and Academic Performance: A Study on Early Childhood and Primary Education (ECPE) Classes. *IAA Journal of Education, 10*(2), 1–12.

Baleni, L. S., Malatji, K. S., & Wadesango, N. (2016). The influence of peer tutoring on students' performance in a South African university. *Journal of Communication, 7*(1), 127–133.

Balfanz, R., & Byrnes, V. (2019). *Early warning indicators and intervention systems: State of the field*. Sage.

Begum, R., Wadood, F., & Hussain, M. (2024). Impact of socio-economic factors on students' academic performance and mental health: Exploring educational challenges and emotional distress. *Pakistan Research Journal of Social Sciences, 3*(1).

Boughey, C. (2010). Understanding teaching and learning at foundation level: A 'critical' imperative. Beyond the university gates: Provision of extended curriculum programmes in South Africa, 4–10.

Boughey, C. (2020). The role of tracking systems in supporting at-risk students in higher education. *South African Journal of Higher Education, 34*(1), 123–137.

Boughey, C. (2022). Not there yet: Knowledge building in educational development ten years on. *Teaching in Higher Education, 27*(8), 992–1004.

Boughey, C. (2024). Conditions constraining the potential of educational development to impact on the transformation of teaching and learning. *Critical Studies in Teaching and Learning (CriSTaL), 12*(2), 44–59.

Cassells, L. (2018). The effectiveness of early identification of at-risk students in higher education institutions. *Assessment & Evaluation in Higher Education, 43*(4), 515–526.

CHE. (2023a). VitalStats: Public and private higher education data 2023. Council on Higher Education.

CHE (2023b). Understanding epistemic access and success of students from historically disadvantaged backgrounds in South African universities [https://www.che.ac.za/sites/default/files/inlinefiles/CHE%20MONITOR%2016_DIGITAL%20COPY_HI%20RES.pdf#:~:text=The%20CHE%20fulfils,findings%20through%20reports%20and%20publications].

CHE (2024). Higher Education Monitor No. 16: [https://www.che.ac.za/sites].

Christiansen, I. M., & Slammert, L. (2006). A multi-faceted approach to research development (II): Supporting communities of practice. *South African Journal of Higher Education, 20*(1), 17–30.

Chiramba, O., & Ndofirepi, E. S. (2023). Access and success in higher education: Disadvantaged students lived experiences beyond funding hurdles at a Metropolitan South African university. *South African Journal of Higher Education, 37*(6), 56–75. https://doi.org/10.20853/37-6-6021

Council on Higher Education. (2020). *"Extended programmes with an integrated foundation phase : Theoretical considerations for curriculum design." CHE no. 14, December 2020*. CHE.

Department of Education. (1997). *Education White Paper 3: A programme for the transformation of higher education*. Government Gazette.

DHET (2012). Policy for the provision of foundation programmes in public higher education. Department of Higher Education and Training.

DHET (2023). Ministerial statement on university funding: 2024/25 to 2026/27. Department of Higher Education and Training.

Ehren, M., & Baxter, J. (2020). *Trust, accountability, and capacity in education system reform: Global perspectives in comparative education*. Routledge.

Emma, O., & Owen, A. (2025). *Integrating data-driven strategies for effective personalized learning: Challenges, implementation, and future trends*. Van Schaick.

Garraway, J., & Bozalek, V. (2019a). Theoretical Frameworks and the extended curriculum programme. *Alternation, 26*(2).

Garraway, J., & Bozalek, V. (2019b). Reconfiguring foundational pedagogies through theoretical frameworks. *Alternation, 26*(2), 1–7.

Johnson, U. (2017). *Success or failure? Student experiences of the extended curriculum programme (ECP)*. Doctoral dissertation, PhD Thesis. Durban: University of Kwa Zulu-Natal.

Kgwete, E. M., & Malatji, K. S. (2021). Problem-solving as a teaching strategy: Promoting active learning in a South African university of technology. *Gender and Behaviour, 19*(2), 18027–18034.

Kolb, J. (1974). *Research methodology: Basic concepts in the methodology of social sciences research* (3rd ed.). Blackwell.

Leibowitz, B., Bozalek, V., Van Schalkwyk, S., & Winberg, C. (2015). Institutional context matters: The professional development of academics as teachers in South African higher education. *Higher Education, 69*(2), 315–330.

Lynam, S., Milani, R., & Lafarge, C. (2024). "Listening to the voices of minoritized ethnic post-graduate researchers in the United Kingdom", *New Vistas, 10*(1). https://doi.org/10.36828/newvistas.250

Malatji, K. S., Makhanikhe, J., Marhaya, L., & Wadesango, N. (2017). Widening of access in South African higher education institutions: Implications for teaching and learning. *PONTE International Journal of Science and Research, 73*(11).

Malatji, K. S., Ramollo, J. K., & Malatji, M. J. (2023). The use of Kolb's theory to conduct effective teaching in mathematics: A case study of learner articulation gap in a grade 3 class. *Journal of Educational Studies, 22*(1), 26–45.

Malatji, K. S., & Singh, R. J. (2018). Implications of the articulation gap between Geography learners in secondary schools and universities. *Alternation, 21*, 91–108.

Mampane, R. (2021). Tracking systems in South African higher education: Enhancing support for underprepared students in Extended Curriculum Programmes. *Journal of Higher Education Research and Development, 40*(3), 58–72.

Mampane, R., & Khoza, M. (2020). The role of academic tracking in supporting the success of ECP students: A case study from a South African university. *Journal of Higher Education Policy and Management, 42*(3), 201–217.

Megbowon, F. K., Palesa, M. K., Bongiwe, K., & Sipokazi, M. (2023). Challenges of first-year extended curriculum programme students at a University in South Africa. *International Journal of Learning, Teaching and Educational Research, 22*(4), 178–194.

Mgutshini, T. (2022). *Extended curriculum programmes: Challenges and opportunities*. Van Schaick.

Marais, K. (2022). *An interpretative phenomenological analysis of academic identity development in part-time lecturers at a private higher education institution in South Africa*. Doctoral dissertation, Stellenbosch: Stellenbosch University.

Mkhize, T., & Balfour, R. (2021). Bridging the gap: The effectiveness of extended curriculum programmes in South African universities. *Journal of Educational Development in Africa, 19*(3), 56–75.

Mokganya, G., & Zitha, I. (2024). Exploring the impact of the extended curriculum programme (ECP) as a crucial motivating landing space: A study on student transition in higher education. *International Journal of Research in Business & Social Science, 13*(10).

Motsoenyane, N. (2020). Embedding academic support in extended curriculum programmes: Lessons from rural institutions. *South African Journal of Education, 40*(2), 44–63.
Mouton, M. (2023). *Exploring the lived experiences of extended curriculum programme students.* Master Dissertation, Stellenbosch: Stellenbosch University.
Mukora, J. (2022). *Equitable access to higher education in South Africa: A policy and practice analysis.* UCT Press.
Mutinta, G. (2022). Mental distress among university students in the Eastern Cape Province. *South Africa. BMC psychology, 10*(1), 204.
Mthethwa, N. E. (2022). *Developing academic self-efficacy in disadvantaged black students from the extended curriculum programmes, Master's thesis.* University of Pretoria.
Ngwenya, P., & Ramphal, R. (2023a). Infrastructure and access: The role of technology in supporting extended curriculum programme students in rural universities. *Journal of Higher Education Research and Innovation, 38*(1), 99–120.
Ngwenya, P., & Ramphal, R. (2023b). Data-driven approaches to student success in South African universities: The role of tracking systems in extended curriculum programmes. *International Journal of Higher Education Research, 45*(2), 58–74.
Ogude, N., Majozi, P., Mathabathe, K., Mthethwa, N. (2021). The management of student success in extended curriculum programmes: A case study of the University of Pretoria's Mamelodi Campus, South Africa. *South African Journal of Higher Education, 35*(4). https://doi.org/10.20853/35-4-4242.
Parés, M. E., Angelats, E., Espín-López, P. F., Navarro, J. A., Munaretto, D., Salvador, J., & Bachmann, J., et al. (2024). IOPES, a new full-fledged approach to provide an end-to-end tracking system for emergency staff. *International Journal of Disaster Risk Reduction,* 10(2), 104-120.
Reddy, T. (2004). *Higher education and social transformation: South Africa case study.* Kagiso.
Scott, I. (2020). The use of tracking systems in higher education: Lessons from extended curriculum programmes in South Africa. *Higher Education Studies, 39*(4), 213–229.
Scott, I. (2009). Academic development in South African higher education. Higher education in South Africa: A scholarly look behind the scenes. *Journal of Educational Studies, 22*(1), 1–21.
Scott, I., Yeld, N., & Hendry, J. (2007). *Case for improving teaching and learning in South African higher education: Higher education monitor No. 6.* Council on Higher Education.
Shalaby, R. A. H., & Agyapong, V. I. (2020). Peer support in mental health: Literature review. *JMIR Mental Health, 7*(6), 155–172.
Shay, S. K., & Wolff, J. (2016). *New generation extended curriculum programmes: Report to the DHET*. University of Cape Town.
Shay, S., Wolff, K., & Clarence-Fincham, J. (2016). New generation extended curriculum programmes: Report to the Department of Higher Education and Training. Department of Higher Education and Training.
Sit, F. (2024). Global inequalities in access and success in higher education. *New Vistas, 10*(1).
UNESCO. (1998). *World declaration on higher education in the twenty- first century-vision and action.* Unesco Publishing.
UNESCO. (2015). *Rethinking education: Towards a global common good*. UNESCO.
United Nations. Cape Town: SDG Hubs. https://academicimpact.un.org/content/sdg-hubs.
Vermunt, J. D., & Donche, V. (2017). A learning patterns perspective on student learning in higher education: State of the art and moving forward. *Educational Psychology Review, 29*, 269–299.

Chapter 6
Proposing a Multidimensional Protocol for Identifying At-Risk University Students

Orifha Sinthumule, Aaron Khosa, Goitsemodimo Molocwa, and Usanda Bekebu

Abstract Identifying and tracking at-risk students is a prevalent practice in higher education institutions worldwide; however, the methodologies employed for this purpose often lack clarity and consistency. This study addresses this issue by developing a three-phase early identification model for at-risk students at the module, programme, and socio-activity levels within South African higher education institutions. To this end, a PRISMA systematic literature review was conducted, analysing 16 studies selected from an initial screening of 100 studies retrieved from the Web of Science database. The final selection of studies examined the methodologies and frameworks for identifying at-risk university students. The study emphasises that module-level interventions should be implemented by teaching practitioners within specific modules. Conversely, programme-level interventions should involve collaborative efforts among teaching practitioners (module lecturers or university educators), programme managers, and institutional teaching and learning support stakeholders. The protocol aims to enhance student retention and academic success by utilising students' module grades and socio-economic factors to identify at-risk students. The study suggests that early identification of risks associated with modules, programmes, and socio-economic status may improve student retention and success in higher education. Follow-up pilot or longitudinal studies are recommended to evaluate the effectiveness of the proposed model in identifying, tracking, and supporting at-risk students.

O. Sinthumule (✉)
Faculty of Humanities, School of Education, University of Limpopo, Polokwane, Mankweng, South Africa
e-mail: orifha.sinthumule@ul.ac.za

A. Khosa · G. Molocwa · U. Bekebu
Extended Curriculum Programme, University of Venda, Thohoyandou, South Africa
e-mail: aaron.khosa@univen.ac.za

G. Molocwa
e-mail: goitsemodimo.molocwa@univen.ac.za

U. Bekebu
e-mail: usanda.bekebu@univen.ac.za

K. Dube and L. M. Tjabane (eds.), *Advancing Sustainable Development Goals in Higher Education*, https://doi.org/10.1007/978-3-032-27646-9_6

Keywords At-risk student · Identification and tracking · Student success · Riks model · PRISMA

6.1 Introduction and Background

Research highlights a burgeoning body of literature on student success, dropout rates, and at-risk students in higher education (Ngulube, 2025; Rahmani et al., 2024; Magd et al., 2022). Simultaneously, issues related to access have spurred the development of various policies and protocols in higher education administration (Universities South Africa Strategic Framework 2020–2026; #FeesMustFall Movement, 2016; Makoelle and Mosito, 2025). International frameworks, such as UNESCO's Education 2030, rooted in Sustainable Development Goal 4 (SDG 4), aim to ensure equitable access to affordable, high-quality tertiary education by 2030 (UNESCO, 2024). Likewise, UNICEF (2023) encourages governments to prioritise public funding for foundational and higher education. All these global initiatives are fundamentally committed to realising a global ideal: increased access to higher-quality education for all segments of society.

In the South African context, a central policy response is the National Student Financial Aid Scheme (NSFAS). According to Wildschut et al., (2018) the NSFAS is a scheme aimed at funding low-income students, thereby ensuring equal university access, and curbing the historical racial imbalances in higher education. Setho (2024) as in Kale (2024) confirm that since 2018, NSFAS has provided comprehensive funding to students from households earning below R350,000 annually, with an additional R3.8 billion allocated in 2024 to support the so-called missing middle (household income ranging between R350,000 and R600,000). Other policy responses include flexible learning programmes and initiatives to increase enrolment among underrepresented groups (Wildschut et al., 2018). Despite enhanced access, there is escalating concern regarding student dropout rates from Higher Education Institutions (HEIs) (Rahmani et al., 2024). Recent global studies affirm that dropout rates remain notably high (Hanson, 2025; Lamichhane et al., 2025). Moreover, contributing factors have expanded beyond financial issues to include outdated curricula, inadequate support services, political interference, and limited aid (Lamichhane et al, 2025).

The massification of education, aimed at democratising learning, has unintentionally contributed to higher dropout rates (James, 2025; Luescher, 2023; Nyagope, 2023). Elevated enrolment numbers may compromise instructional quality due to strained learning environments, limited infrastructure, and diverse student motivations (Mukhanji et al., 2016). South Africa faces particularly low academic achievement and high dropout rates (Oberholster, 2024). To curb this disastrous trend, institutions can implement early at-risk identification and tracking systems to mitigate the gap between enrolments and dropout rates. Stressors such as accommodation issues, financial difficulties, and academic demands hinder progression (Mdletshe, 2025). Moreover, Cele et al. (2025) suggest that the high dropout rate poses a significant challenge to breaking the cycle of poverty, affecting the socio-economic status of many residents. Previous studies by Letseka and Maile (2008) and Muswede (2017) found a first-year dropout rate of approximately 30%, with nearly 50% of students discontinuing within their first three years. Recent estimates by Fraser (2024) suggest that the first-year dropout rate may now reach 60%. Cele et al. (2025) and Netanda (2024) expound on the reason for this attrition and attribute it to contributing factors such as late NSFAS payments, inadequate academic preparation from secondary school, social integration problems, and exposure to racism at historically white institutions, and institutional factors, such as delayed lecturer feedback and resource constraints.

The COVID-19 pandemic exacerbated these trends, with notable increases in dropout from the third year onwards in 2020, particularly at historically disadvantaged rural universities (Branson et al., 2023). University rules, such as the N+ rule, which limits funding for students who exceed the standard programme duration, further increase dropout risks, as students facing academic challenges risk academic exclusion and funding loss (Minister of Higher Education, Science and Technology 2021). High dropout rates pose a significant challenge to achieving Sustainable Development Goals (SDGs), particularly SDG 1 (ending poverty) and SDG 4 (inclusive and equitable quality education). These objectives are inherently interconnected, as education serves as a pivotal means of eradicating poverty CSDG (2024). The identification of at-risk students and the implementation of targeted interventions are the only factors that contribute to achieving these global objectives (Dube et al., 2024).

The relationship between socio-economic status (SES) and academic performance warrants critical attention in the literature. Students from diverse pre-tertiary and socio-economic backgrounds who enter the same educational environment may exacerbate existing inequalities in teaching and learning processes (Cassells, 2018). While some research indicates a moderate positive correlation between SES and academic achievement (Munir et al., 2023; Selvitopu & Kaya, 2023), other studies contend that in higher education, this relationship remains positive but weak, with prior academic performance and university experience serving as more robust predictors (Rodriguez-Hernandez et al., 2020). Despite this, Nekhubvi et al. (2025) and Munir et al. (2023) argue that students from low-SES backgrounds are disproportionately at risk of dropping out or failing, often due to a lack of supplementary academic resources, such as private tutoring, textbooks, and technological tools. Furthermore, the intricacies of low-income students' experiences considerably influence their academic success and psychological well-being within South African higher education institutions (Bantjes et al., 2020; Walker, 2020). However, low SES cannot be generalised as the sole reason for high dropout rates, especially given financial aid schemes such as the NSFAS (Nekhubvi et al., 2025; Edge Education, 2025).

A significant challenge is that educational expansion has not mitigated disparities in academic outcomes between students of high and low socio-economic status (Lui & Ng, 2023). Therefore, higher education institutions (HEIs) must develop programmes and pedagogical interventions explicitly designed to address this gap and promote academic success for all students. Institutional discourse often associates at-risk students with performance metrics, success, and throughput rates. For instance, early warning systems designed to identify at-risk students have been shown to be effective in improving academic performance (Cassells, 2018). Additionally, Nekhubvi et al. (2025) emphasise that research explicitly links socio-economic backgrounds to indicators of academic achievement. According to Behr et al. (2020), McDermott et al. (2019), empirical studies on student withdrawal are generally conducted from two perspectives: the individual perspective, which emphasises student attitudes, behaviours, and academic performance, including prior learning, and the institutional perspective, which focuses on contextual factors related to families, schools, and communities.

To identify at-risk students, researchers commonly utilise predictors and models that incorporate demographic information, socio-economic variables, academic preparation, and institutional performance metrics (Brown et al., 2023; Cardona et al., 2020; Yan et al., 2025). The present study seeks to develop a protocol for the scholarship of teaching and learning that identifies and monitors at-risk students primarily through academic grades and socio-economic factors at the module and programme levels. A subsequent investigation will assess the validity, reliability, and effectiveness of the proposed protocol.

6.2 Literature Review

The phrase "At-risk" is an operational concept used in most institutions of higher learning to refer to students who require academic remediation or social intervention to progress in their studies (Deil-Amen, 2011). The at-risk concept, as discussed by Castro (2014) and Dix et al. (2020), refers to a student's preparedness deficit, often resulting from socio-economic and cultural factors beyond the student's control. This study addresses the gap in the literature by drafting a model protocol to aid in the identification and tracking of at-risk students.

6.2.1 Theoretical Framework

Care Theory (Noddings, 2002) was employed. As argued by Atif et al. (2020), delivering timely interventions requires teaching staff to identify at-risk students and act on that information to motivate and support them. The theory, originating in the 1980s (Gilligan, 1982; Noddings, 2002), centres on the relational nature of care. Noddings (2012) conceptualises this as a cyclical relationship: the cared-for (the student) indicates a need, the carer (the practitioner) responds with available resources, and the cared-for acknowledges the care received. O'Brien (2010) explains that academics who intentionally establish and nurture such caring relations benefit both their students and themselves, notwithstanding various difficulties.

This study aims to enhance the understanding of the phenomenon of at-risk students among academics and higher education professionals by developing and applying a protocol for their identification. This is essential, as failing to identify at-risk students can be costly for students and institutional retention rates. Hampered academic progress, such as failing a module, can delay degree completion, incur additional costs, reduce self-confidence, prompt a change of major, lead to dropouts, and ultimately hinder active participation in the global economy (Smith et al., 2012; Sosibo & Katiya, 2015). Despite this, the relationship between care and academics accumulated experience has not been extensively explored, particularly in the context of identifying and tracking at-risk students in higher education. The personal academic journeys of teaching practitioners may inform a caring approach, shaping how they view and support students through the lens of care theory.

Studies examining various factors influencing poor students' academic outcomes and attrition have highlighted the importance of reducing attrition by identifying at-risk students early (Goradia & Bugarcic, 2019). This implies that teaching practitioners have the onus to identify and track students who risk failing the module. The earlier the identification and tracking are performed, the better the intervention and measures taken to respond to the risk. Sinthumule and Zitha (2024) contend that students' self-determination and proactivity are significant contributors to academic success. However, this study argues that, while self-determined, students may still face significant risks due to factors beyond their control. This aligns with Sinthumule and Zitha's (2024) observation that factors other than self-determination can lead to academic risk and potential dropouts. The existing literature identifies numerous contributing factors. Lizarte-Simon and Gijón Puerta (2022) emphasise factors such as the integration of academic and social life, degree commitment, collegiate stress, institutional advising, counselling, motivation, financial strain, and academic efficacy. For analytical clarity, this study groups these contributing factors into two primary categories: academic and socio-economic, which are discussed in the following sections.

6.2.2 Academic Factors

Some academic challenges faced by university students can be traced back to before they entered university. As Sosibo and Katiya (2015) suggest, lowering matriculation standards led students to struggle academically as they transitioned from high school to higher education. The more significant challenge this poses is that many pupils will pass their matric results, many of whom might not qualify to register at a university. Those who meet university requirements may not meet the programme's minimum requirements, which can lead to lower academic performance in their first year. Meeting the minimum enrolment requirements for a programme is suitable for entry into institutions of higher learning. Boughey (2020) highlighted that students who meet the bare minimum "are at risk of failure because of the poor quality of their previous educational experiences."

Meeting the minimum academic requirements for a programme does not automatically mean that a student can handle the academic rigour required to succeed

(Sinthumule & Zitha, 2024). While ECPs provide a "bridge" to combat persistent inequities, they also provide a preparatory phase that helps first-year students catch up on key skills such as literacy, numeracy, critical thinking, and study techniques (Brits et al., 2024). Furthermore, Brits et al. (2024) stressed that the ECP assumption is that the programme provides adequate time to complete introductory university courses, potentially increasing student success at the individual level and reducing the dropout rates at the institutional level. Some academic determinants of at-risk students include poor attendance and low grades in summative assessments, first within a particular module and then across modules in the programme itself (Kemp et al., 2023). A study by Muresherwa and Bama (2024) showed a problematic relationship between irregular or non-attendance of classes and poor academic performance. However, it is worth stressing that class attendance also implies active participation in class, including the module content, lecturer, and fellow students.

Student under-preparedness exacerbates risk factors, particularly in institutions of higher learning. Students enter institutions of higher learning underprepared, with little prior recognition of learning. Tucker and McKnight (2019) argued that initial pre-admissions assessments across programmes are a feasible predictor of academic success, as they help identify at-risk students before academic enrolment. Similar to some programmes in South Africa, some require the National Benchmarking Test (NBT) as a prerequisite. In contrast, others use admission assessments to determine students' eligibility for admission.

The learning environment has also been identified as an academic indicator of students' success or risk. According to Jardinez and Natividad (2024), a supportive and inclusive learning environment can significantly mitigate students' barriers, fostering academic success, personal growth, and resilience. Conversely, a poorly designed or unsupportive learning environment exacerbates students' difficulties, leading to disengagement, poor performance, or even higher dropout rates at the university. Factors such as access to technology, quiet study spaces (e.g. libraries), and academic support services are foundational for promoting fair opportunities for university students (Ullah & Usman, 2023). Equally important are intangible elements such as a sense of belonging, mentorship, and a non-judgmental environment that promotes self-confidence and motivation. Additionally, at-risk students regularly benefit from environments that are responsive to their unique needs. According to Müller et al. (2023), this includes flexibility in course delivery, such as blended or online learning options, personalised academic interventions, and programmes that address emotional well-being. Institutions that prioritise inclusivity and adapt their strategies to accommodate diverse students' experiences are more likely to encourage these students to persist and succeed in their studies. By doing so, universities can enhance retention rates and contribute to the broader goal of equity and access to education.

According to Halali et al (2023), language barriers pose a significant challenge for students who are non-native speakers of the institution's prescribed medium of instruction, affecting various aspects of their academic journey. These barriers can hamper understanding of course material, limit effective communication with lecturers and peers, and ultimately affect overall academic performance negatively.

The inability to fully understand lectures, assignments, or exams due to language difficulties can create frustration and division, further compounding students' struggles (Halali et al., 2023). Beyond academic challenges, language barriers can affect students' social integration within the university community. Non-native speakers may struggle to participate in discussions, group projects, or extracurricular activities, which are essential for building networks and promoting a sense of belonging (Civan & Coskun, 2016). This social isolation can lead to diminished confidence, reduced motivation, and increased stress, negatively affecting emotional well-being and academic persistence. Ucheh (2024) claimed that language barriers can limit access to support services and resources. Students may need help managing institutional processes, such as understanding administrative requirements, seeking academic help, or utilising campus facilities. Universities must adopt a proactive and diversified approach to address these barriers.

6.2.3 Socio-Economic Factors

Socio-economic factors, as a protocol for recognising at-risk students, necessitate the analysis of key indicators that consider students' social and economic factors that may impact their well-being and academic progress. According to UWC (2019) of socio-economic issues exacerbate the risk of students to not pass their classes; these include financial struggles, food insecurity, lack of healthcare, mental health issues, poor off-campus living conditions, and high commuting costs. These socio-economic factors encompass family income, occupation, parental education, and social status, which collectively foster student access to opportunities, well-being, and resources (Sarra et al., 2019). It is important to proactively identify at-risk students early to shape interventions that can enhance academic performance (Kumari et al., 2024; Sarra et al., 2019).

Generally, students from high-income socio-demographic backgrounds are expected to have better access to educational resources, technology, and extracurricular opportunities. Meanwhile, lower-income students experience financial constraints that limit their ability to engage fully in their studies (Chanda, 2024). Hence, a study by the Nekhubvi et al., (2025) concluded that financial situations, as a socio-economic factor, directly correlate with student success. Some students are financially burdened by sharing their allocated bursaries with their families, which adds to their academic pressure and the risk of academic failure. This was also highlighted by Munir et al. (2023), who stressed that students' socio-economic position affects academic performance, with those in a higher socio-economic status performing better academically than those in a lower socio-economic status.

Chanda (2024) highlighted the social status and occupation of students' family members, parents, or siblings. Such a family background could influence academic progress and the first-hand support they may receive at home regarding their educational, social, and future career opportunities. Universities cannot accommodate 100% of their enrolled students, and higher social-class families can better understand

and afford the needs of off-campus students. The challenge of staying off campus can be unsafe accommodations, often far from campus, leading to absenteeism, as students may not have enough financial resources to commute to and from campus.

Although universities provide various resources and support services, students negotiate their academic journeys. The support may not be sufficient, and it is challenging to cater to all students, especially those willing to use such services. These support services aim to foster holistic development by providing social, academic and practical support (Council of Graduate Schools, 2022). Social support networks, such as cultural groups and social organisations, foster a sense of community and belonging, enabling cultural awareness and interpersonal connections (Chanda, 2024). Academic support involves tutoring, workshops, and advisory services to enhance progression. Emotional support services, such as student counselling, are significant for dealing with the psychological well-being of students, assisting them in dealing with anxiety, stress, and various mental health concerns (Chanda, 2024). However, practical support services, such as financial aid schemes, career counselling, and accommodation for differently abled students, enhance equitable access for all students (Pascarella & Terenzini, 2005).

The quality of teaching and learning conditions, resource availability, and department-student relationships significantly impact progression (Chanda, 2024). Students tend to feel shy about communicating this to the lecturer; therefore, lecturers must provide support to such students. This can take the form of online classes and asynchronous study materials available to students. Therefore, a hybrid teaching method must be adopted to benefit all students, irrespective of their socio-economic status (Nekhubvi et al., 2025). Furthermore, a student's gender must also be considered when lecturers use a holistic approach to managing at-risk students (MacMaoilir & McGillicuddy, 2023). Male and female students might be dealing with different gender-based challenges in their attempts to achieve academic success. While lecturers cannot be overly sympathetic, they must be aware of and well-informed about the university's policies and practices designed to assist students better. Module lecturers or programme managers may refer students to student counselling and other relevant offices as a means of social intervention when a student's academic risk is identified. Lecturers and programme managers must be aware of the services available at the university in the first instance so that they can refer their students to the most appropriate interventions and support.

Students' academic preparedness and the availability of institutional support initiatives are significant determinants of progression, as at-risk students lacking such support are more likely to falter in their studies. This underscores the importance, as noted by Kumari et al. (2024), of proactively identifying risk elements that may impact student progression, making early detection a priority. Consequently, academic staff must be equipped to navigate the broader socio-economic and administrative challenges students face. This includes advising students on institutional procedures for applying for assessment concessions or special examinations, as well as guiding those who need to temporarily suspend their studies, a process commonly termed placing studies in abeyance. Although lecturers may not directly administer these processes, they play a crucial role in providing initial guidance and referrals.

For instance, a lecturer may be unable to complete financial forms for a student with outstanding debt. However, they are positioned to bring critical documents, such as an Acknowledgement of Debt (AOD) form, to the student's attention and refer them to the relevant administrative stakeholders, thereby facilitating access to essential support pathways.

6.3 Research Method

A systematic literature review (SLR) was conducted following the Preferred Reporting Items for Systematic Reviews and Meta-Analyses (PRISMA) guidelines to identify, summarise, and evaluate the current methodological approaches for identifying at-risk students in higher education. The review's scope was structured using the PICO framework (Population, Intervention, Comparison, Outcome). The population comprised students in a university's extended curriculum programme. Intervention refers to tools, models, or methodological approaches for identifying and tracking at-risk students. A comparison is not directly applicable because this is a scoping review of existing approaches. This study focuses on academic grades and early determinants, tailored to a less-resourced, financially constrained institution where a significant portion of students rely on government assistance programmes such as NSFAS. The intended outcome is the proposal of an operationalised, effective protocol for identifying at-risk students

A comprehensive search was conducted in the Web of Science Core Collection. The search strategy employed the following terms in the title, abstract, or keywords: at-risk student, student at risk, identifying students at risk, or a model to identify a student at risk. The search adhered to five eligibility criteria: (1) date: articles published between 2003 and 2025; (2) language: English only; (3) publication type: research articles, book chapters, and conference proceedings; (4) geography: no restrictions; and (5) study quality: publications in higher education journals or special issues focussed on higher education, including teaching and learning. The initial search returned 100 records. Metadata (title, abstract, keywords, authors, journal, and year) were exported to a Microsoft Excel spreadsheet for screening. Two researchers independently screened all records based on their titles and abstracts. Inconsistencies were resolved through discussion until a consensus was reached. After screening, 84 studies were excluded. Exclusions were based on (a) scope misalignment with the review objective (identifying at-risk student methodologies), (b) insufficient study quality as per the predefined criterion, or (c) proposed methodologies requiring significant financial resources or complex integration with institutional Learning Management Systems (LMS), thus lacking cost-effectiveness for broad application.

A final set of 16 studies was selected for the full-text review and synthesis. The systematic selection process is detailed in the PRISMA flow diagram (Fig. 6.1). The database search was last updated on 20 March 2025. The synthesis focussed specifically on the intervention, namely the tools and methodological frameworks used for identification, as operationalised in the selected SLRs relevant to the study's objective.

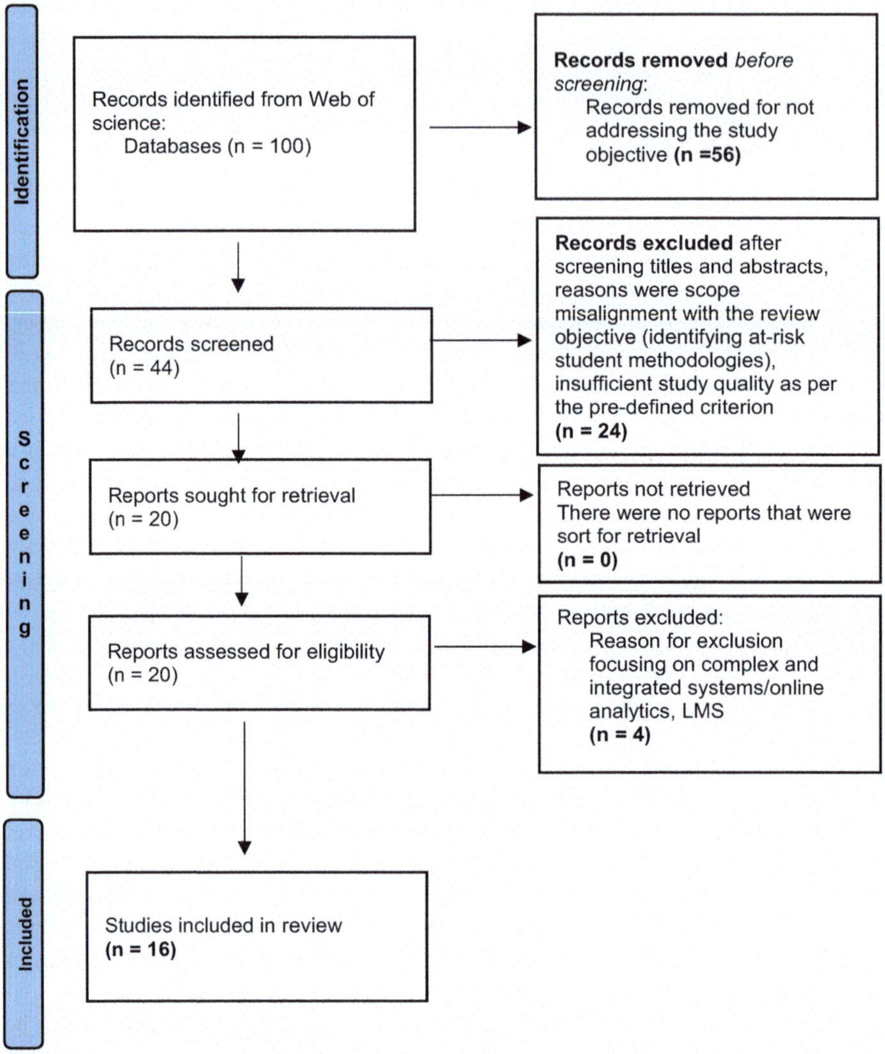

Fig. 6.1 PRISMA 2009 flow diagram

6.4 Results and Discussion

Phlegar and Rose (1988) argued that identifying students potentially at risk of not completing their programmes is key to prevention. This aligns with Lui and Ng (2023), who emphasised that identifying students at risk of failing or dropping out is essential for instructional remediation, retention, and success. Early identification enables timely intervention to aid in programme completion. This study structures the identification protocol into three interrelated levels: (i) the module level, (ii) the programme level, and (iii) socio-economic factors. These levels acknowledge the direct relationship between a department offering a module and the module's role within the broader programme. For the module and programme levels, a qualitative approach was used to collect data through student consultations, lecturer observations, and feedback. These dialogues were essential for identifying the root causes of low grades in the subject. The third level focuses on socio-economic factors as risk enablers. Here, dialogues explore reasons for poor performance beyond academic pressure, ensuring that risk is assessed through both grades and broader socio-economic contexts.

Figure 6.2 shows the levels and their relationships from which at-risk student factors may be identified. This includes identification at the module and programme levels. Moreover, the identification of socio-economic factors that may lead to risk at the module or programme level is necessary.

The following section examines the identification of at-risk students across three levels: first, at the module level, then at the programme level, and finally through the lens of socio-economic factors that may contribute to academic risk.

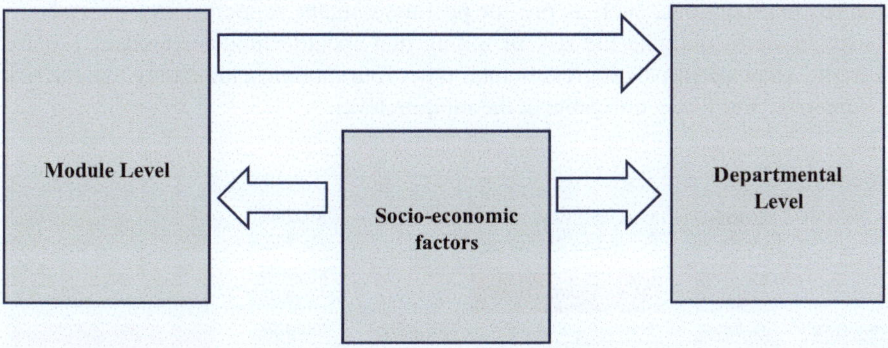

Fig. 6.2 Identification and tracking of at-risk students model

6.4.1 Identification and Tracking of At-Risk Students at a Module Level

An analysis of the 16 PRISMA-included studies indicates that while Learning Management Systems (LMS) aid identification, the focus remains on academic grades and early detection of socio-economic factors. Consequently, this study developed complementary models for module and departmental implementation that align with the variable relationships shown in Fig. 6.1. In this phase, the course convener identifies at-risk students within a specific module based on the results of formal and informal learning activities. These indicators may include attendance, online engagement, and assessment performance. The module's teaching practitioner is responsible for assessing, identifying, and tracking these students. Guided by the theory of care, practitioners should also monitor students' performance across other programme modules to gather more comprehensive evidence. The following model was used for module-level identification.

Figures 6.3 present student 1A and student 1B, the module assessment results for which the two students are registered, and the academic performance for each registered module. Lastly, this figure shows Student A's academic performance for each enrolled student in the programme.

Figures 6.3 above identify students at risk of failing a specific module. Students 1A and 1B are progressing satisfactorily in five concurrent modules but have each failed one assessment: Student 1A in Module B and Student 1B in Module C. The lecturer should identify this module-level risk upon receiving the assessment results. Through the lens of care theory, lecturers can intervene directly or refer students to relevant institutional support units. The at-risk status necessitates tracking and deploying targeted interventions, such as pre- or post-assessment support, extra tutorials, or consultations to mitigate the risk of failing that specific module. Student 1A, for example, shows positive progress in their other four modules, indicating that the risk is contained and likely mitigable at the module level.

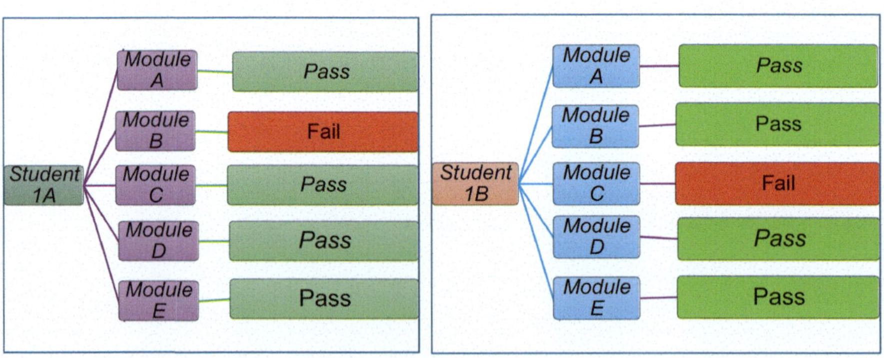

Fig. 6.3 Identification of at-risk students at the module level

6.4.2 *Identification and Tracking of At-Risk Students at the Programme Level*

At the departmental level, the focus extends beyond the individual modules. Students at risk of failing multiple modules are consequently at risk of failing their first-year programmes. Data for this level are aggregated from empirical identification and tracking already conducted by module practitioners. Once module results are finalised, they must be cascaded to the project for students who are at risk of being dismissed. This level explicitly identifies students at risk of failing across several modules within the same programme.

Subsequent Figs. 6.4 illustrate the identification and tracking of at-risk students at the programme level with Module A and B enrolled by a number of students.

Figure 6.4 identifies at-risk students at the programme level. Students 1A, 1C and 1D failed an assessment in Module A. Moreover the same students, except student 1E, but with student 1D, failed an assessment in Module B. This pattern indicates unsatisfactory academic progress, as students are failing multiple modules at the same programme level, elevating the issue from a module-specific concern to a programme-level problem. The core challenge is that students are at a high risk of failing multiple modules in a single semester. This risk, initially identified at the module level, must be escalated to the programme level for a comprehensive assessment. Here, interventions requiring specialist referrals beyond the lecturer's expertise may be prescribed. While lecturers can mitigate risks within their module content, their primary focus is on teaching and student mastery. Programme-level risks require broader and coordinated support. Nevertheless, special departmental attention is required for Student 1A and 1C, who are at risk of failing the two modules. However there is a moderate risk of student 1E and 1D who are likely to fail one module if no intervention are taken sooner by the teaching practitioners.

Lecturers consolidate teaching and learning activity data, flagging Student 1B as at-risk for Modules A and C at the departmental level. Intervention and tracking at this level involve the lecturer and the Head of Department (HOD) and focus on the

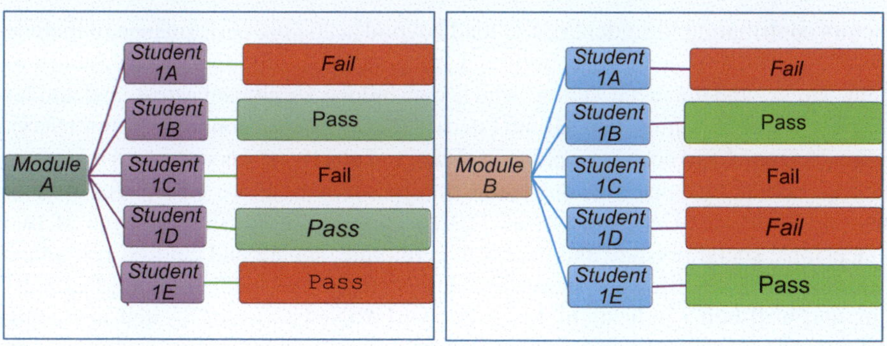

Fig. 6.4 Identification and tracking of at-risk students at the programme level

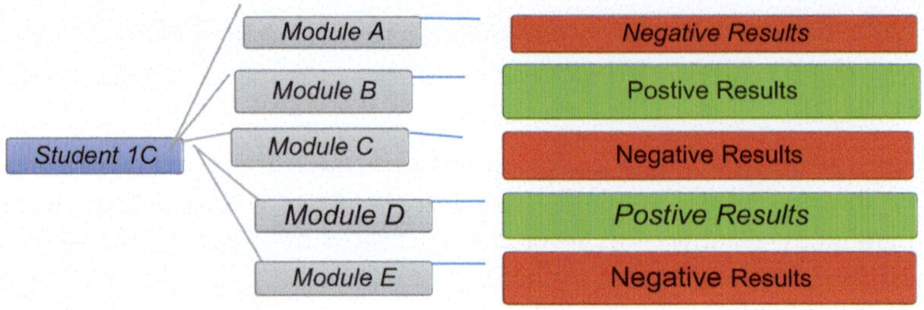

Fig. 6.5 Identification of at-risk Students at the programme 1 level

student's risk within the programme. The student's risk is considered medium and likely mitigable at this level.

Figure 6.5 presents the departmental-level identification model, highlighting the case of Student 1C. Enrolled in five Extended Curriculum Programme (ECP) modules, the student demonstrates problematic progress and is at high risk of failing three modules (A, C, and E), which constitute 60% of their course load. This departmentally identified risk directly threatens institutional pass and throughput rates, necessitating immediate intervention and referral to relevant specialists.

The figure above shows that special attention is needed for Student 1C at the departmental level and across all the modules. In this scenario, lecturers must consolidate the results of all the formal and informal activities that Student 1C undertakes. This helps track and intervene in the student's risk. This intervention is conducted at the departmental level by the lecturer and the HOD, with the aim of tracking students' risk of failing modules and teaching and learning activities within the programme. Student 1C is at a high risk of failing and dropping out of the ECP.

6.4.3 Identification and Tracking of At-Risk Students: Socio-Economic Factors

Factors contributing to student risk extend beyond academic performance to include social determinants such as institutional integration, family background, commitment, stress, and financial status, which can induce significant stress and hinder progress (Lizarte-Simon and Gijón Puerta, 2022). Socio-economic status (SES), defined as social and economic position (Mistry et al., 2010), encompasses resources such as family income, education, and occupation, which are critical for student success. Rodríguez-Hernández et al. (2020) conceptualised SES in terms of two dimensions: social prestige and access to economic, social, and cultural resources. This aligns with the lifestyle perspective, in which higher wealth, occupational prestige, and well-being indicate higher SES (Bird & Newport, 2017, cited in Zungu, 2022).

In South Africa, although the National Student Financial Aid Scheme (NSFAS) has enhanced access to higher education, insufficient funding continues to leave many students without support, thereby exacerbising academic stress alongside socio-economic stress (SES) (Ravhuhali et al., 2022). Students from low socio-economic status (SES) backgrounds frequently lack critical resources, which heightens their vulnerability to academic failure and dropout; however, the higher education literature does not establish a direct link between SES and academic performance (Rodríguez-Hernández et al., 2020). Drawing on frameworks from basic education, the South African Policy on Screening, Identification, Assessment, and Support (SIAS) advocates tiered levels of support, as demonstrated in previous studies by Pillay and Tlale (2024). This study associates these support levels with academic risk statuses: students who attain passing marks are classified as low risk; those who miss or fail assessments are deemed moderate risk and require intervention. At the module level, personalised consultations between lecturers and students are essential for diagnosis.

Lecturers are advised to be mindful of their authority, and tutors or peer mentors can serve as crucial intermediaries for students who are more comfortable with their peers. If a lecturer lacks the training to provide broader support, referral to teaching and learning support services via the academic head of department (HOD) is required with the student's consent. This mitigates the risk of module failure caused by socio-economic factors. The HOD should track referrals and note contextual challenges, such as group assignments that disproportionately burden low-SES students who may lack funds, time, or stable Internet (UNICEF, 2013). Effective intervention requires a caring, individualised approach, as students need to feel "respected, understood, appreciated, and cared for to promote growth and learning" (Owusu-Ansah and Kyei-Blankson, 2016, p. 1).

Mentoring is particularly effective for students from low-income families (Abeysinghe & Fernando, 2024). However, socio-economic stress can manifest as hopelessness and depression (Ravhuhali et al., 2022), necessitating a swift referral to professional support. Ultimately, the success of an intervention depends on its implementation. Support structures are often under-resourced, with staff grappling with widespread student poverty (Nyewe, 2018). Overwhelmed units may refer students to overcrowded public services, potentially leading to discontinued support and dropouts. Therefore, lecturers, HODs, and support staff must systematically track the implementation and progress of the interventions.

6.5 Conclusion and Recommendation

This study proposes a multidimensional protocol for the early and continuous identification of students at risk of module or programme failure, academic exclusion, or dropout. The protocol is designed as a practical tool for teaching practitioners, academic managers, and support stakeholders to implement targeted module-level interventions. Its application aims to address high dropout rates by enabling timely remedial support, thereby mitigating academic risks and improving the outcomes. This approach is grounded in care theory, where practitioners proactively respond to assessed student needs with appropriate resources and referrals to enhance student well-being and reduce long-term institutional costs.

The operational implementation commences with the initiation of teaching activities. Academic risk is identified through key indicators, including engagement and performance metrics, such as assessment grades, LMS interaction levels, group participation, and consultation attendance. While students may be reluctant to disclose socio-economic or personal difficulties, they generally establish trust with at least one academic practitioner. This relationship is essential for enabling consent-based referrals to institutional or external support services, which can effectively address the underlying barriers to success. However, adopting this protocol may encounter several challenges. These include a lack of comprehensive policy frameworks, inconsistent availability of counselling or financial aid, difficulties in standardising data collection across diverse modules and programmes, an increased workload for academic staff, the potential for student stigmatisation due to being "flagged," and variability in the reliability of self-reported socio-economic data.

The study recommends integrating the identification protocol directly into institutional technological ecosystems, such as the central Learning Management System (LMS) and student record software. While the existing literature often proposes advanced machine learning techniques (e.g. SMOTE, logical regression) for this purpose (Abeysinhe & Fernando, 2024; Singh & Alhulail, 2022), such solutions can be financially prohibitive. In contrast, the proposed protocol is cost-effective and centres on structured practitioner engagement. Institutions are advised to embed it within their LMS to generate early alerts at the module, departmental, and faculty levels. Concurrent support interventions should include academic tutoring, enhanced advising, peer mentoring programmes, and structured social engagement initiatives. The South African Department of Higher Education should consider a policy-mandated pilot study to evaluate the protocol's effectiveness in reducing attrition. Finally, this study recommends that future research develop and test a scalable, algorithm-enhanced tracking system within institutional management platforms, focussing on validating the protocol's efficacy in real-world settings.

Ethical Considerations This protocol is proposed to identify at-risk students; no ethical considerations are required for this study.

References

Abeysinghe, D. V. D. S., & Fernando, M. S. D., (2024). An integrated framework for an educational early warning system with mentor matching. *IAFOR Journal of Education, 12*(3), 243–271. https://files.eric.ed.gov/fulltext/EJ1453801.pdf

Atif, A., Richards, D., Liu, D., & Bilgin, A. A. (2020). Perceived benefits and barriers of a prototype early alert system to detect engagement and support 'at-risk' students: The teacher perspective. *Computers & Education, 156,* 103954. https://doi.org/10.1016/j.compedu.2020.103954

Bantjes, J., Saal, W., Lochner, C., Roos, J., Auerbach, R. P., Mortier, P., & Stein, D. J. (2020). Inequality in mental healthcare utilisation among first-year university students in South Africa. *International Journal of Mental Health Systems, 14*(1), 5. https://doi.org/10.1186/s13033-020-0339-y

Behr, A., Giese, M., Teguim Kamdjou, H. D., & Theune, K. (2020). Dropping out ofuniversity: A literature review. *Review of Education, 8*(2), 6. https://doi.org/10.1002/rev3.3202

Bird, R., & Newport, F. (2017). What determines how Americans perceive their social class?. *Gallup Polling Matters.* https://www.studocu.com/en-us/document/university-of-colorado-boulder/principles-of-microeconomics/what-determines-how-americans-perceive-their-social-class/51999972

Boughey, C. (2020). Council of higher education: Extended programmes with an integrated foundation phase: Theoretical considerations for curriculum design. Briefly Speaking, (14), 1–16. https://www.che.ac.za/publications/monitoring/brieflyspeaking-extended-programmes-integrated-foundation-phase-theoretical

Branson, N., Ranchhod, V., & Whitelaw, E. (2023). *South African student retention during 2020: Evidence from system wide institutional data: Southern Africa Labour and Development Research Unit*. University of Cape Town.

Brits, G. P., Eloff, I., & Van Staden, S. (2024). Using extended curriculum programmes to improve student success at universities. *International Journal of African Higher Education, 11*(2), 58–83. https://www.ajol.info/index.php/ijahe/article/view/296914

Brown, D. A., Soto, J. G., Anand, S., Weimer, N. E., & Black, V. A. (2023). Risk factors and effectiveness of implemented academic interventions on student retention at a hispanic-serving institution. *Journal of College Student Retention: Research, Theory & Practice,* 25(2), 398–419. https://doi.org/10.1177/152102512098697810.1177/152102512098698

Cassells, L. (2018). The effectiveness of early identification of 'at risk' students in highereducation institutions. *Assessment & Evaluation in Higher Education, 43*(4), Article 515526. https://doi.org/10.1080/02602938.2017.1376033

Castro, E. L. (2014). Underprepared and "At-Risk": Disrupting deficit discourses inundergraduate STEM recruitment and retention programming. *Journal of StudentAffairs Research and Practice, 51*(4), 407–419. https://doi.org/10.1515/jsarp-2014-0041

Cele, S. M. K., Pietersen, D., & Gaillard, C. (2025). Black African students' social and academic identities in South African Universities Vis-À-Vis student dropout: A social justice and philosophical perspective. *Journal of Culture and Values in Education, 8*(1), 240–251. https://culturandvalues.org/

Center for Sustainable Development Goals, Universitas Brawijaya. (2024). *Education and poverty: Interconnected issues in the global consciousness*. https://sdgs.ub.ac.id/education-and-poverty-interconnected-issues-in-the-global-consciousness/

Chanda, T. M. (2024). Student retention in higher learning institutions of Zambia. *International Journal of Research Publication and Reviews, 5*(6), 433–441. https://www.researchgate.net/publication/381120196_Student_Retention_in_Higher_Learning_Institutions_of_Zambia

Civan, A., & Coskun, A. (2016). The effect of the medium of instruction language on theacademic success of university students. *Educational Sciences: Theory andPractice, 16*(6), 1981–2004. https://doi.org/10.12738/estp.2016.6.0052

Council of Graduate Schools. (2022). Degree Completion. https://cgsnet.org/graduate-student-retention-and-completion

Deil-Amen, R. (2011). Socio-Academic integrative moments: Rethinking academic and social integration among two-year college students in career-related programs. *The Journalof Higher Education, 82*(1), 54–91. https://doi.org/10.1080/00221546.2011.11779085

Dix, N., Lail, A., Birnbaum, M., & Paris, J. (2020). Exploring the "At-Risk" student label through the perspectives of higher education professionals. *The Qualitative Report, 25*(11), 3830–3846. https://nsuworks.nova.edu/tqr/vol25/iss11/4

Dube, K., Booysen, R. M., & Chilli, M. (2024). Redefining education and development: Innovative approaches in the era of sustainable goals. In *Redefining education and development: Innovative approaches in the era of the sustainable development goals* (pp. 1–8). Springer Nature Switzerland. https://doi.org/10.1007/978-3-031-69954-2_1

Edge Education. (2025). The Way Forward. https://url-shortener.me/82KQ

Fraser, L. (2024). University dropout crisis for South Africa. Business Tech. Lifestyle. https://businesstech.co.za/news/lifestyle/771036/university-dropout-crisis-for-southafrica/

Gilligan, C. (1982). New maps of development: New visions of maturity. *American Journal of Orthopsychiatry, 52*(2), 199. https://doi.org/10.1111/j.1939-0025.1982.tb02682.x

Goradia, T., & Bugarcic, A. (2019). Exploration and evaluation of the tools used to identifyfirst-year at-risk students in health science courses: A systematic review. *Advances InIntegrative Medicine, 6*(4), 143–150. https://doi.org/10.1016/j.aimed.2018.11.003

Halali, A. A. S., Ismail, L., Samad, A. A., Razali, A. B., & Noordin, N. (2023). Mediation: Effects of language anxiety and prior learning experience on academic speaking: Challenges and strategies among libyan students in Malaysian universities. *World Journal of English Language, 13*(6). https://doi.org/10.5430/wjel.v13n6p385

Hanson, M. (2025). College dropout rates. Education Data Initiative. https://educationdata.org/college-dropout-rates

James, A. (2025). *Experts Explain Gains in an 'Inefficient' HE System.* University World News: African Edition. https://www.universityworldnews.com/post.php?story=20250923205015630

Jardinez, M. J., & Natividad, L. R. (2024). The advantages and challenges of inclusive education: striving for equity in the classroom. *Shanlax International Journal of Education, 12*(2), 57–65. https://orcid.org/0000-0002-5438-0884

Kale, M. G. (2024). *What now? Experiences of everyday unemployment among Black graduates in South Africa* (Doctoral dissertation, University of Oxford).

Kemp, P. R., Bradshaw, J. M., Pandya, B., Davies, D., Morrell, M. J., & Sam, A. H. (2023). The validity of engagement and feedback assessments: Identifying students at risk of failing. *BMC Medical Education, 23*(1), 866. https://doi.org/10.1186/s12909-023-04828-7

Kumari, R., Satyam, S. M., Bairy, L. K., Rehman, A., Shetty, P., & Mohammed, C. A. (2024). Crossing horizons: unravelling perspectives on enhancing medical students' success through at-risk factor exploration. *BMC Medical Education, 24*(1), 835. https://doi.org/10.1186/s12909-024-05819-y

Lamichhane, B. D., Giri, O. P, Singh, B. D., Kafle, B. J, & Lama, P. B. (2025). Determinant factors of University students' dropouts: A qualitative analysis. *International Journal of Economics and Management Studies, 12*(4), 35–45. https://doi.org/10.14445/23939125/IJEMS-V12I4P105

Letseka, M., & Maile, S. (2008). *High university dropout rates: A threat to South Africa's future.* Human Sciences Research Council. https://repository.hsrc.ac.za/handle/20.500.11910/4967

Lizarte Simón, E. J., & Gijón Puerta, J. (2022). Prediction of early dropout in higher education using the SCPQ. *Cogent Psychology, 9*(1), 2123588. https://doi.org/10.1080/23311908.2022.2123588

Luescher, T. (2023). *Report on transformation at SA universities: It is about more than numbers.* HSRC: Equitable Education and Economies. Report on transformation at SA universities: it is about more than numbers.

Lui, A. K. F., & Ng, S. C. (2023). Looking through the fog of remote Zoom teaching: A casestudy of at-risk student prediction. *International Journal of Mobile Learning andOrganisation, 17*(4), 499–516. https://doi.org/10.1504/IJMLO.2023.133753

MacMaoilir, E., & McGillicuddy, D. (2023). I'm actually a female empowerer: Student perspectives on a critical pedagogical approach to re/engage at-risk females in school. *Education and Urban Society, 55*(9), 1047–1069. https://doi.org/10.1177/00131245221106725.

Magd, H., Thirumalaisamy, R., & Ruzive, B. (2022). Identification of at-risk students and higher education institutions' strategies for rising their potential: A conceptual approach. *Global Business and Management Research, 14*(2s), 118–133.

Makoelle, T. M., & Mosito, C. P. (2025). Introduction: Towards an Inclusive Higher Education.

McDermott, E. R., Donlan, A. E., & Zaff, J. F. (2019). Why do students drop out? Turning points and long-term experiences. *The Journal of Educational Research, 112*(2), Article 270282. https://doi.org/10.1080/00220671.2018.1517296

Mdletshe, L. N. (2025). A study of the students' struggles for housing at the university of KwaZulu-Natal. In *Campus student activism, politics, and citizenship in the 21st Century*. IGI Global Scientific Publishing.

Minister of Higher Education, Science and Technology. (2021). Guidelines for the department of higher education and training bursary scheme for students at publicUniversities. DHET. 1–20. https://url-shortener.me/82HN

Mistry, R. S., Benner, A. D., Biesanz, J. C., Clark, S. L., & Howes, C. (2010). Family and social risk, and parental investments during the early childhood years as predictors of low-income children's school readiness outcomes. *Early Childhood ResearchQuarterly, 25*(4), 432–449. https://doi.org/10.1016/j.ecresq.2010.01.002

Mukhanji, J. M, Ndiku, J. M, & Obaki, S. (2016). Effect of increased student enrolment on teaching and learning resources in Maseno University, Kenya. *The International Journal of Social Sciences and Humanities Invention*, 3(3), 2349–2031. https://url-shortener.me/82H7

Müller, C., Mildenberger, T., & Steingruber, D. (2023). Learning effectiveness of a flexiblelearning study programme in a blended learning design: Why are somecourses moreeffective than others? *International Journal of Educational Technologyin Higher Education, 20*(1), 10. https://doi.org/10.1186/s41239-022-00379-x

Munir, J., Faiza, M., Jamal, B., Daud, S., & Iqbal, K. (2023). The impact of socio-economic status on academic achievement. *Journal of Social Sciences Review, 3*(2), 695–705. https://doi.org/10.54183/jssr.v3i2.308

Muresherwa, G., & Bama, H, K. (2024). Towards improving class attendance in higher education: A case study of a university of technology in South Africa. *International Journal of Learning, Teaching and Educational Research, 23*(5), 19–439. https://doi.org/10.26803/ijlter.23.5.22

Muswede, T. (2017). Colonial legacies and the decolonisation discourse in post-apartheid South Africa: A reflective analysis of student activism in higher education. *African Journal of Public Affairs, 9*(5), 200–210. http://hdl.handle.net/2263/59082

Nekhubvi, V., Sinthumule, O., Zitha, I., Mukhodobwane, R., & Mulaudzi, T. (2025). Effects of students' socioeconomic status on academic performance: A case of first-year students at a South African University. *International Journal of Business Ecosystem & Strategy (2687–2293), 7*(2), 414–421. https://doi.org/10.36096/ijbes.v7i2.758

Netanda, S. R. (2024). Race, age, and gender as attributes of student attrition in an open distance e-learning (ODeL) landscape. *Commonwealth Youth Development, 21*(1), 1–16. https://doi.org/10.25159/2663-6549/14517

Ngulube, P. (2025). Predicting academic success and identifying at-risk students: A systematic review of data analytics and machine learning approaches in higher education institutions. *Educational Administration: Theory and Practice, 31*(1), 117–134.

Noddings, N. (2012). The caring relation in teaching. *Oxford Review of Education, 38*(6), Article 771781. https://doi.org/10.1080/03054985.2012.745047

Noddings, N., (2002). *Educating moral people: A caring alternative to character education.* Teachers College Press. ERIC—ED468125—educating moral people: a caring alternative to character education, 2002.

Nyagope, T. S. (2023). Massification at higher education institutions: Challenges associated with teaching large classes and how it impacts the quality of teaching and learning in South Africa. *International Journal of Environmental, Sustainability and Social Science, 5*(1), 209–220. https://doi.org/10.32478/al-mudarris.v6i2.1842

Nyewe, P. (2018). Pre-Service educators' perceptions on strategies employed by their mentor teachers to foster inclusive education in their classrooms. In *ICERI2018 Proceedings*, pp. 4620–4622. Nyewe_P_Edu_2018.pdf

O'Brien, L. M. (2010). Caring in the ivory tower. *Teaching in Higher Education, 15*(1), Article 109115. https://doi.org/10.1080/13562510903487818

Oberholster, J. (2024). The future of commerce education: Facilitating first-year student success in commerce faculties in South Africa. In E. Loots, & J. Oberholster (Eds.), *Rethinking commerce education in South Africa: The case for change to develop future fit business leaders*, AOSIS Books, Cape Town, 19–31. https://doi.org/10.4102/aosis

Owusu-Ansah, A., & Kyei-Blankson. (2016). Going back to the basics: Demonstrating care, connectedness, and a pedagogy of relationship in education. *World Journal of Education, 6*(3), 1–9. http://wje.sciedupress.com/

Pascarella, E., & Terenzini. (2005). Students' experiences in college, 2: A third decadeof research. Jossey-Bass. https://url-shortener.me/82E7

Phlegar, J. M., & Rose, R. M. (1988). At-Risk students: Approaches to identification and intervention. The Rhode Island Department of Education. https://files.eric.ed.gov/fulltext/ED297230.pdf

Pillay, E., & Tlale, L. D. K. (2024). The implementation of the screening, identification, assessment and support policy to manage and support teaching and learning processes in kwazulu-natal. *E-Journal of Humanities, Arts and Social Sciences, 5*(13). https://doi.org/10.38159/ehass.20245132

Rahmani, A. M., Groot, W., & Rahmani, H. (2024). Dropout in online higher education: Asystematic literature review. *International Journal of Educational Technology inHigher Education, 21*(1), 19. https://doi.org/10.1186/s41239-024-00450-9

Ravhuhali, F., Nendauni, L., Sikhwari, T. D., Lavhelani, N. P., Dama, N. G., Matodzi, T., & Manyage, T. (2022). A self-determination theoretical approach to survival strategies of on-campus and off-campus students from low-income families. *South African Journal of Higher Education, 36*(3), 231–248. https://doi.org/10.20853/36-3-4658

Rodríguez-Hernández, C. F., Cascallar, E., & Kyndt, E. (2020). Socio-economic status and academic performance in higher education: A systematic review. *Educational Research Review, 29*, 100305.

Rodríguez-Hernández, C. F., Cascallar, E., & Kyndt, E. (2020). Socio-Economic status andacademic performance in higher education: A systematic review. *EducationalResearchReview, 29*, Article 100305. https://doi.org/10.1016/j.edurev.2019.100305

Sarra, A., Fontanella, L., & Di Zio, S. (2019). Identifying Students at risk of academic failure within the educational data mining framework. *Social Indicators Research, 146*, 41–60. https://doi.org/10.1007/s11205-018-1901-8

Selvitopu, A., & Kaya, M. (2023). A meta-analytic review of the effect of socioeconomic status on academic performance. *Journal of Education, 203*(4), 768–780. https://doi.org/10.1177/00220574211031978

Setho, L. (2024, January 14). *Govt unveils R3.8 billion loan scheme for 'Missing Middle' students*. SABC News. https://www.sabcnews.com/sabcnews/910418-2/ Accessed June 10, 2026.

Singh, H. P., & Alhulail, H. N. (2022). Predicting student-teachers' dropout risk and early identification: A four-step logistic regression approach. *IEEE Access, 10*, 6470–6482. https://doi.org/10.1109/ACCESS.2022.3141992

Sinthumule, O., & Zitha, I. (2024). From high-school sciences to university accounting in South Africa: How prior learning affects students' success. *International Journal of Research in Business & Social Science, 13*(6). https://doi.org/10.20525/ijrbs.v13i6.3571

Smith, G., Therry, L., & Whale, J. (2012). *Developing A Model for Identifying Students at Riskof Failure in a First-Year Accounting Unit.* https://doi.org/10.5539/hes.v2n4p91

Sosibo, Z., & Katiya, M. (2015). Closing the loop between access and success: Early identification of at-risk students and monitoring as key strategies used by a South African University. *International Journal of Educational Sciences, 8*(2), 271–279. https://url-shortener.me/82CM

Tucker, L., & McKnight, O. (2019). Assessing the validity of college success indicators for the at-risk student: Toward developing a best-practice model. *Journal of College Student Retention: Research, Theory & Practice, 21*(2), 166–183. https://doi.org/10.1177/1521025117696822

Ucheh, M. S. (2024). Communication barriers and coping mechanisms: A closer look at international students' struggles. https://doi.org/10.13140/RG.2.2.22666.44484

Ullah, A. & Usman, M. (2023). Role of libraries in ensuring quality education at higher education institutions: A perspective of Pakistan. *Inverge Journal of Social Sciences, 2*(4), 13–22. https://doi.org/10.63544/ijss.v2i4.57

UNESCO. (2024). *Tranforming education towards SDG 4: Report of a global survey on country actions to transform education.* UNESCO. https://www.unesco.org/en/articles/transforming-education-towards-sdg4-report-global-survey-country-actionstransform-education Accessed June 10, 2026.

UNICEF. (2013). WASH in schools empowers girls' education: Proceedings from the menstrual hygiene management in schools virtual conference 2012. https://url-shortener.me/82AK

UNICEF. (2023). Transforming education with equitable financing. United Nations Plaza.

University of Western Cape. (2019). UWC student retention and success framework document. https://url-shortener.me/82A7

Walker, M. (2020). The well-being of South African University students from low-income households. *Oxford Development Studies, 48*(1), 56–69. https://doi.org/10.1080/13600818.2019.1672143

Wildschut, A., Mncwango, B., Rogan, M., Rust, J., Fongwa, S., & Meiring, L. (2018). The National Student Financial Aid Scheme (NSFAS) and its impact: Exploring the absorption into employment of NSFAS-funded graduates.

Yan, X., Fan, J., Levine, R. A., Jacobs, E., & Stronach, J. (2025). Analysis of risk factors for student retention. *Journal of College Student Retention: Research, Theory & Practice*. https://doi.org/10.1177/152102512513784

Zungu, B. P. (2022). Socio-Economic and academic challenges faced by black undergraduatestudents from disadvantaged backgrounds at the wits school of education. [Dissertation]. *University of the Witwatersrand*. https://url-shortener.me/829O

Chapter 7
Assessing Student Success Through Alumni Achievements in Visual Arts and Design

Zanele Lucia Radebe

Abstract Globally, traditional metrics such as self-efficacy, academic achievement, completion, retention, and persistence, which are often used to evaluate student success during university enrollment, fail to capture the comprehensive impact of educational programs on students. This study measures student outcomes by documenting the success stories of alumni from the Visual Arts and Design (VAD) department at the University of Technology (UoT) in South Africa. Utilizing a qualitative, interpretative approach, this research examines the career trajectories of three alumni: a fashion designer, independent curator, and practicing artist. Using document and thematic analyses, this study examined the factors contributing to alumni success. Braun and Clarke (Qualitative Research in Psychology 3:77–101, 2006) six-step thematic analysis was used to sort the data qualitatively into emerging themes. This study found that alumni success is not only dependent on technical skills provided by the UoT but also on factors such as professional mentorship, industry partnership, competitions, social media, and artistic consistency for long-term sustainable careers. Thus, the study advocates for moving beyond traditional metrics of measuring student success to alumni, as they have a distinct perspective on the effectiveness of their education in their careers. This study highlights the need for HEIs to integrate alumni feedback into curriculum design. This study calls for the revision of the VAD curriculum at the UoT to ensure that it better prepares students for industry demands and long-term sustainable careers.

Keywords Student success · Alumni achievements · Visual arts and design

Z. L. Radebe (✉)
Department of Visual Arts and Design, Faculty of Human Sciences, Vaal University of Technology, Vanderbijlpark, South Africa
e-mail: zaneler@vut.ac.za

K. Dube and L. M. Tjabane (eds.), *Advancing Sustainable Development Goals in Higher Education*, https://doi.org/10.1007/978-3-032-27646-9_7

7.1 Introduction

In higher educational institutions, traditional metrics have long been used to measure student success (e.g., York et al., 2015). These metrics include self-efficacy, academic achievement, completion, retention, and persistence (Soika, 2021). While these metrics are significant for measuring students' success, they fail to capture the long-term impact of institutional curricula (Salazar & Schelbe, 2021). These metrics are not as effective, especially in the Visual Arts and Design field, where a successful artistic career is determined by how much an artist exhibits their work, the influence their work has on people, and selling art to earn a living (Petrides & Fernandes, 2020).

Several scholars contend that student success should be measured through alumni accomplishments rather than focusing solely on in-university traditional metrics, as alumni achievements give a true reflection of the impact of the university's curriculum post-graduation (Blankstein & Wolff-Eisenberg, 2020; Matthew, 2023; Salazar & Schelbe, 2021; Spence, 2019). The careers of VAD graduates are shaped mainly by freelance work, self-employment, and partnerships with local and global brands in a dynamic and changing industry (Ibáñez-Sánchez et al., 2022). Traditional metrics, especially at UoTs, overlook these aspects and focus on academic performance and degree completion rates (Saunders-Smits & de Graaff, 2012).

The UoT mandate prioritizes practical learning, industry relevance, and professional preparedness (Chinomona et al., 2013). While conventional universities focus mainly on critical theoretical knowledge, UoTs are expected to produce students who are ready for employment and entrepreneurial endeavors (Ncube & Lekhanya, 2025). However, a critical gap exists when student success is measured solely through in-university years rather than assessing graduates in the industry with long-term, sustainable career achievements.

While the UoT aims to produce industry-ready graduates, there is a paucity of comprehensive tracking of graduate outcomes in terms of employment rate, career advancement, and industry impact (Enihe & Omopariola, 2023; Pico-Saltos et al., 2022). Many graduates struggle to get employment and start businesses in their field of study, despite having received training through university programs (Graham et al., 2019; Yende, 2021). This raises the issue of the effectiveness of the UoT mandate. In addition, the evolving industry and the mismatch between the university curriculum and industry demands make it difficult for graduates to transition from student life to sustainable careers. Few studies have explored alumni success in Visual Arts and Design at UoTs in South Africa, highlighting a significant gap in evaluating long-term student success. Thus, there is a need for a comprehensive method to assess the effectiveness of the UoT curriculum beyond traditional metrics.

Therefore, this chapter shifts away from viewing student success through traditional metrics to alumni outcomes by examining the career trajectories of three alumni graduates: a fashion designer, an independent curator, and a practicing artist from the VAD department at the UoT in South Africa. Hence, this study aims to measure student outcomes by documenting the success stories of alumni from the Visual Arts and Design Department at the University of Technology in South Africa. The primary research question for this study is as follows: How do alumni achievements reveal gaps and strengths in the VAD curriculum at a South African UoT? The objectives of this chapter are to identify critical factors contributing to their success and to provide recommendations for enhancing university programs to better prepare future students for successful, sustainable careers.

This chapter provides a nuanced way in which UoTs can improve the VAD curriculum. This will be achieved by focusing on the factors that play a significant role in the alumni's success to better prepare graduates for sustainable careers in the art industry. This aligns with Agenda 2030 for Sustainable Development, particularly Sustainable Development Goal (SDG) 4: Quality Education. SDG 4 focuses on inclusive, equitable, and quality education that emphasizes lifelong learning opportunities for all individuals. By analyzing alumni success, this study provides crucial data on how the UoT curriculum can prepare students for lifelong learning opportunities. This will not only prepare students for employment but also further their skills development and adaptability throughout their professional lives. Examining the factors contributing to alumni success will also provide critical information for advancing inclusive and equitable education. This means that graduates from diverse backgrounds can also obtain meaningful and sustainable career pathways as a result of high-quality education.

The chapter is organized as follows: First, it presents a literature review on different conceptualizations of student success and critiques traditional metrics in higher education institutions. Next, it discusses the research design employed in this study. The findings of this study are presented and discussed. Finally, the study concludes and provides recommendations.

7.2 Literature Review

This section presents literature on different conceptualizations of student success and critiques traditional metrics in higher educational institutions.

7.2.1 Student Success in Higher Education: Beyond Traditional Metrics

York et al. (2015) define student success as "academic achievement, engagement in educationally purposeful activities, satisfaction, acquisition of desired knowledge, skills and competencies, persistence, attainment of educational outcomes, and post-college performance". These academic achievements are ostensibly acquired through traditional measures such as graduation rates, retention, and post-graduation employment to determine "success" (Blankstein & Wolff-Eisenberg, 2020). Early studies conceptualized student success by focusing on grades and retention (Kuh et al., 2006). The following were identified as crucial elements to defining student success by Kuh et al., "(a) academic achievement, (b) persistence, (c) engagement in educationally purposeful activities, (d) acquisition of desired educational outcomes of knowledge, skills, and competencies, (e) persistence, (f) satisfaction, and (g) post-college performance".

In contrast, Soika (2021) emphasized the variability of traditional metrics based on who is measuring. Soika (2021) noted that although there are numerous methodologies to measure students, there are common metrics used in higher education, including self-efficacy, academic achievement, completion, retention, and persistence. Self-efficacy denotes how much effort an individual is willing to put forth to achieve their goals (Bandura, 1977). Self-efficacy is related to one's capability to learn new knowledge and acquire skills that can be used to achieve the goals that one sets for oneself (Matovu, 2020). DiBenedetto and Schunk (2022) define academic self-efficacy as "a dynamic motivational belief that influences the goals we set, how hard we persist, and the amount of effort we employ". Thus, self-efficacy is related to one's ability to complete a task.

In the university setting, academic achievement correlates with performance outcomes which indicate how well a student has fulfilled the objectives set out for the activity (Steinmayr et al., 2014). Schneider and Preckel (2017) argue that stimulating meaningful learning through clear presentation of material, student-centered activities, and conceptually challenging assignments is also highly linked to academic achievement. York et al. (2015) highlight that grades and grade point average (GPA) are extensively utilized measures of scholarly accomplishment in higher education. Kool et al. (2018) further assert that academic ability serves as a moderator for academic achievements. Therefore, there are different criteria used in higher education to measure academic achievement, however, what they have in common is that they all measure the intellectual capacity of an individual (Steinmayr et al., 2014).

The term completion describes students who obtain their degree in a specific allocated time, normally six years after enrolling in post-secondary education for the first time. The objective of the completion rate is to ensure that the enrollment rate is at the same pace as the degree completion rate (Dawson et al., 2020). Retention is defined as "a measure of the rate at which students persist in their educational program at an institution, expressed in a percentage" (McCollum, 2018). Therefore, retention is related to the percentage of students who return to the course from one year to the next. Conversely, persistence signifies the rate of student progression from one benchmark to the next, for example, a student who enrolls from semester one to semester two (Soika, 2021).

Traditional metrics, such as self-efficacy, academic achievement, completion, retention, and persistence, have been conventionally employed to measure students' success while they are enrolled in the university. However, numerous studies contend that higher education should not refine student success to a simplistic, traditional metric measure that is only based on grades and academic accomplishments (Salazar & Schelbe, 2021; Spence, 2019). Rather, they should also focus on post-graduation employment to define "success" (Blankstein & Wolff-Eisenberg, 2020).

Mathews (2023) corroborates this by stating that to track student success, universities should shift their focus to graduates after they finish their university programs.

In higher education, research on student success through alumni focuses on various facets, such as financial implications for student success, which include funding scholarships and financial contributions from alumni (Dean & Jendzurski, 2021); engagement through mentorship with higher education as they assist in bridging the gap between academic preparation and professional demands (Bieger, 2024; Pico-Saltos et al., 2022; Saunders-Smits & Graaff, 2012), as well as engagement curriculum development (Basri & Sitorus, 2023; Nanda et al., 2021).

7.2.2 The Role of Alumni in Improving the Curriculum

Barron (2015) asserts that the most effective way to measure a program's success is through its alumni, as they are a true reflection of the program. Matthew (2023) further emphasizes that alumni involvement in higher education can offer value in their social and professional capacities as mentors to students. As representatives of their alma mater, alumni are significant for the enrollment of students and raising funds (Dean & Jendzurski, 2021). Nanda et al. (2021) posits that one way to provide appropriate education is by involving alumni. Alumni engagement provides higher educational institutions with information on how they can transform academic and support programs to ensure student engagement and success. Furthermore, alumni play a key role as input givers in curriculum development, which aims to address community needs and demands (Basri & Sitorus, 2023).

Universities should reform their curriculum through alumni, as they have applied the knowledge they acquired from the university program in society (Basri & Sitorus, 2023). For example, a study conducted at the Tshwane University of Technology on arts alumni revealed a high unemployment rate attributed to curriculum misalignment with industry requirements (Yende & Mugovhani, 2021). The critical gap identified by alumni inspired the development of a transformed curriculum. The revised curriculum now includes skills that align with industry requirements, aiming to enhance graduates' employability and entrepreneurship skills (Yende, 2021). This finding aligns with Saunders-Smits and Graaff (2012), who state that alumni are significant for curriculum evaluation and development as they have a distinct perspective to assess how well their education assisted them in their careers. In addition to highlighting the flaws they encountered while in university, they can also share insights from their daily job experiences regarding the information and abilities students currently need to thrive in their professional careers.

7.2.3 *Role of Alumni in Preparing Students for Sustainable Successful Careers*

Alumni are significant for the growth and development of higher education institutions. At the UoT, alumni are crucial in equipping students for prosperous careers by serving as role models and sources of inspiration (Tulankar & Grampurohit, 2020). Alumni contribute significantly through their professional experience, including financial management, self-discipline, and character formation. Such interactions between alumni and students are crucial as they inspire students to cultivate self-esteem and self-motivation (Kwarteng & Obeng-Ofori, 2021). Alumni's relationship with the university also presents students with invaluable internships and entry-level employment opportunities, whereas some alumni give students from their university first preference for employment opportunities (Balotescu, 2024). For example, Durban University of Technology is actively investing in creating a robust alumni mentorship ecosystem, by utilizing its graduates' knowledge and networks to empower its current students (Peters, 2021).

7.3 Research Design

This study employed a qualitative, interpretative research approach. Oranga and Matere (2023) contend that qualitative research provides a comprehensive examination of phenomena through non-numerical data. This approach was deemed appropriate for the study's objectives, as it enabled the researcher to concentrate on a thorough analysis of the three alumni to gain an in-depth understanding of the factors that contributed to their success. A biographical research design was employed, given its focus on documenting another individual's life and experiences (Creswell & Poth, 2016). This methodology enabled the researcher to understand the lives and career development of alumni through their personal and professional milestones.

The population included VAD alumni who graduated after 1998 from the UoT in South Africa. To be selected as participants for the study, the alumni were required to have exhibited their work locally and internationally, possess a minimum of ten years' experience in the creative field, and have an active career. This criterion was used to represent a robust view of postgraduate success. Three participants were purposefully selected as the focus of this study: a fashion designer, an independent curator, and a practicing artist. This sample size was based on the principles of data saturation and depth of analysis. Saturation ensured a thorough understanding of the success factors, where additional data did not yield new insights. The small sample size allowed for an in-depth analysis of each participant's career, yielding rich and detailed narratives.

Non-empirical data were collected through a comprehensive document analysis. Documents comprised scholarly journal articles, news articles, research studies, and exhibition catalogues to obtain rich information about the selected alumni. To conduct data collection, the first step was to collect documents on the selected alumni using Google Scholar as a primary search engine, and the search was done using their professional names. The second step involved collecting web page documents on the three selected alumni, using their names as search terms on Google to retrieve the web page documents. The inclusion of webpage documents was based on alumni being renowned people in the creative industry, who have been profiled across diverse platforms, including news, creative, entertainment, and academic sites. The third step involved systematically saving the collected documents in three distinct folders (fashion designer, independent curator, and practicing designer).

To ensure the integrity and quality of the data, documents were validated through rigorous cross-referencing involving the consultation of multiple sources. Source triangulation, a technique in which three or more sources converge on the same information, was employed. This method assisted in establishing the accuracy and reliability of these data.

Creswell (2014) asserts that conducting research ethically includes being honest about the use of data and protecting the privacy and confidentiality of research participants. In this study, the confidentiality and anonymity of the participants' data were carefully adhered to throughout the study. Furthermore, intellectual property was duly acknowledged, with all information gathered from journal articles, news articles, research studies, and exhibition catalogues appropriately cited and referenced.

Braun and Clarke's (2006) six-step thematic analysis was employed to systematically analyses the non-empirical data collected from the documents on the selected alumni. These steps comprised: (i) becoming familiar with the data; (ii) generating initial codes; (iii) searching for themes; (iv) reviewing themes; (v) defining themes; and (vi) write-up.

To become acquainted with the data, the researcher read and re-read the documents. This step was crucial for identifying data that aligned with the research questions (Byrne, 2022). Patterns in the data were highlighted manually to find the biographical information, achievements, and success factors of each selected alumnus. Codes were meticulously documented for each participant, and concurrent notes were taken to capture initial insights and for the analytical process. The codes that emerged are listed in Table 7.1.

Following the initial coding, the coded data underwent a rigorous review and analytical process to identify codes that shared semantic meanings to form themes (Byrne, 2022). Six themes emerged from the review process: (1) Brief biographical information of the alumni; (2) Professional mentoring: The influence of expert guidance on achieving career success, this theme was derived from codes such as "mentorship programs" "curatorships and mentorship programs", "selected for residencies" observed across the three participants narratives; (3) Collaboration: How alliances enhance creative and professional development theme emerged from codes such as "partnering with big brands" and "working with branded artist" observed across the three participants narratives; (4) The use of competitions in opening up professional opportunities; (5) Social media: Using online presence to increase influence and professional visibility, similarly the theme was derived from codes such as "post going viral on social media" and "virtual exhibitions" and; (6) Consistency

Table 7.1 Coded data

	Fashion designer	Independent curator	Practicing artist
Biographical information	Born in 1975 Obtained a degree in 1995 –More than 20 years in the creative industry	Born in 1980 Obtained a degree in 2005 –More than 12 years in the creative industry	Born in 1992 Obtained a degree in 2014 –More than 10 years in the creative industry
Achievements	–Winning competition –Running a successful mentorship –Showcasing locally and internationally –Receive an honorary doctorate –Receiving fashion awards	–Curating shows locally and internationally –Winning best curated stand internationally –Partnerships with branded artists	–Winning competition –Exhibiting works locally and internationally –Selected to participate in France Olympics (2024) –Collaborating with big brands
Codes identified as success factors	–Winning the competition in 2003 –Brand building –Partnering with brands Use of social media –Receiving funding –Showcasing locally and internationally –Donations –Artistic freedom –Hard work, persistence, and consistency –Virtual exhibitions –Program Re-inventing your style Branding	–Volunteering –Attending –Curatorships and mentorship programs –Selected for residencies –Collaborating with branded artists –Community engagements –Winning competition—best exhibition space—2019	–Mentorship programs –Winning the competition in 2017 –Post going viral on social media –Working with big brands –Weing selected to be showcased on international platforms –Being part of the Olympics in France (2024) –Consistency in producing artifacts Branding

is essential for long-term success through perseverance and commitment. The next step was to define themes and support emerging findings with existing literature to achieve the transferability of the study's conclusion (Guest et al., 2012).

7.4 Data Presentation and Discussion

This section presents brief biographies of the three alumni, outlining their backgrounds, career trajectories, and notable achievements within their art careers.

7.4.1 Brief Biographical Information of the Alumni

The fashion designer (b.1975) creates high-end fashion full of glamour and flair (Jennings, 2015). He received a degree in fashion from the UoT in 1995. An important turning point in his career occurred when he won a competition that inspired him to launch his fashion brand (Brand South Africa, 2013). His career trajectory includes early influences from the UoT that shaped his technical and creative abilities. He also leveraged industry connections and innovations to establish an international brand. He has achieved a lot in the fashion industry, and some of the highlights of his achievements include showcasing his work locally and internationally (Brand South Africa, 2025), receiving fashion awards, receiving an honorary doctorate from another UoT in South Africa, and running a successful mentorship program (Odero, 2020).

The independent curator (b.1980) obtained a degree from the UoT in 2005. Throughout her career, she has contributed to numerous exhibitions, working with the Department of Arts and various community initiatives (Independent Curator International, 2013). She holds a master's degree in curatorial practice. Her career trajectory includes studies in art history and curation, which enabled her to explore independent curation. Her ability to connect with diverse artistic communities has led to the curation of major exhibitions. Her achievements in curatorship practice include winning the best curated stands in prestigious art fairs in the world (Yiba, 2021), working with big brand artists (Nkosi, 2021), and being selected as a judge in prestigious Fine Arts Awards (Kempton Park Express, 2018).

The practicing artist (b.1992) attained a degree at the UoT in 2014, then she furthered her education and obtained a qualification in printmaking in 2015 (Latitudes, 2024). African hairstyling is her theme of exploration (Radebe, 2019). Her career trajectory includes the university's emphasis on individual creativity which fostered her experimentation with new mediums, such as hair. She is well known locally and internationally for her unique artistic signature of using synthetic hair to create a body of work. Her achievements include winning competitions, partnering with established brands, showcasing her work nationally and internationally (Artist Proof Studio, 2025), and being selected as a South African art representative at the 2024 Olympic Pavilion in Paris, France (Digital Drum, 2024).

7.4.2 The Power of Expert Guidance in Shaping Career Success

The first common theme that emerged as a success factor for the three alumni was professional mentorship. The renowned designer initiated a mentoring program to teach emerging fashion designers about the intricacies of the art business. Forbes Africa (2019) noted that the fashion designer's mentorship program was inspired by the lack of guidance regarding the business of fashion.

This gap has also been identified by scholars who assert that there is a gap between higher educational preparedness and industry requirements (Riebe & Jackson, 2014; Woodside, 2018). Zhuang and Tao (2024) underscore the critical importance of alumni inputs, as they provide authentic perceptions of how programs at higher education institutions relate to industry demands through their direct academic and professional encounters.

The significance of mentorship programs is also evident as a success factor for renowned independent curators. In 2013, she was selected to participate in a curatorial practice mentorship program that aimed to provide a platform that created a space for networking and collaboration among upcoming curators (Triangle Network Projects, 2013). The mentorship program was vital in addressing the gap between academic readiness and career expectations (Bieger, 2024) by passing information, experience, and confidence from one generation of curators to another, simultaneously strengthening both artists (Brynjolson, 2017). Currently, this specific type of curatorial mentorship program is not offered to students who want to embark on the path of becoming a curator at the UoT from which the alumni graduated.

Professional mentorship is also a success factor for practicing artists. In 2014, after graduating with a B-tech degree in Fine Arts, she enrolled in a print-making studio where she received her first mentorship (Radebe, 2019). In 2015, she was sent for a four-month residency in the United States of America (USA), where she received her second mentorship abroad (Latitudes, 2024). Finally, in 2013, she was selected for an eight-month residency, where she was fortunate to be mentored by the late legendary art practitioner (Dzimbili, 2019). Her distinctive artistic style was influenced by her work with renowned artists through mentorship programs in South Africa and abroad (Latitudes, 2024). Papadopoulos et al. (2011) concur that mentorship programs assist in fostering personal development, by providing mentees with an in-depth perspective of the industry.

Similar to the fashion designer, Glamour South Africa (2024) noted that the practicing artist learned about the business of art when she furthered her studies at a printing studio after graduating from the UoT.

The finding that alumni sought external mentors due to a lack of business training after graduating highlights a critical gap in industry preparedness in the UoT's curriculum. Such encounters support the argument made in the literature by Saunders-Smits and Graaff (2012), who assert that alumni are significant for curriculum evaluation and development, as they have a distinct position in assessing the impact of their university education on their professional careers.

7.4.3 Alliances, Creativity and Professional Development

Another common theme identified as a significant success factor for three alumni was collaboration with major brands. The concept of collaboration with major brands is becoming increasingly prevalent in the creative field (Ibáñez-Sánchez et al., 2022). Such partnerships offer mutual benefits, enabling both brands and artists to reinvent their brands and attract a larger, more desirable demographic and customer base (Mukhari, 2021). Accordingly, these joint ventures can significantly impact both artists and brands' development and success.

For example, a fashion designer has collaborated with notable brands, such as food, coffee, and clothing brands. These brands secured sponsorships for alumni. Although the fashion designer's brand focuses on high-end fashion designs, these collaborations have facilitated the creation of affordable designs (Keteyi, 2022). Thus, collaboration plays a crucial role in reaching new target markets.

A major factor in the independent curator's success has been collaborating with big brand artists which contributed to her success as a curator (Memani, 2019). Collaborations with big brand artists intensified her credibility by providing her with access to broader audiences, industry networks, and mentorship opportunities. Additionally, the independent curator has collaborated with local artists as part of community engagement (Yiba, 2021).

The practicing artist has also collaborated with big brands in the hair industry, clothing labels, alcohol brands, and jewelry brands (Artist Proof Studio, 2025). These collaborations have given her brand great exposure, increased sales, and opened new opportunities.

7.4.4 The Use of Competitions in Opening up Professional Opportunities

The third common factor contributing to the success of the three alumni is their participation in and achievements in competitions. The fashion designer's career began immediately after he won a talent competition in 2003. On the other hand, the curator won the best curated stand at the most reputable international art fairs, and this great achievement elevated her curatorial career within a global context (Memani, 2019).

Competitions are a success factor for the practicing artist's career. She was selected for a 12-month residency in 2016, which increased her exposure to the brand. The residency exposed her work to potential buyers, exhibition opportunities, and publications. She won 1st prize in a Jewelry Art Fair competition in 2017, which elevated her career to the next level (Arora, 2022). Herrera-Limones et al. (2020) emphasized the importance of competitive accomplishments in boosting artists' chances of obtaining professional success at the international level.

For all three alumni, competitions proved beneficial, but their impact was different. The fashion designer's win paved the way for him to launch a successful brand from the beginning. In contrast, for both practicing artists and independent curators, competitions primarily provide expanded opportunities, recognition, and exposure to their existing brands.

7.4.5 *Online Presence and Its Influence on Professional Visibility*

The fourth success factor that emerged as significant for the alumni is that of social media. The significance of social media is evident in the successful career of fashion designers. During the National lockdown necessitated by the global pandemic, the designer leveraged social media to launch his new collection in 2020. The pre-recorded virtual display functioned as an innovative means to showcase craftsmanship and creativity on digital platforms during the frustrating era of the global pandemic. Although virtual content existed before COVID-19, the pandemic compelled audiences to engage with artistic works through digital media.

In 2019, the practicing artist's artwork gained widespread virality after being posted on social media platforms. This artwork accumulated over 40,000 shares. Nony (2019) describes the artwork as "a piece that used synthetic hair on canvas. Long single lines of hair and pencil traces give shape to female figures seen from the back, at an angle that offers only a glimpse of their faces. Three canvases are placed in a pyramidal shape and linked together via meters of hair twisted in braids that unify and tie these women up in a formation".

Artists leverage digital platforms as tools to enhance their visibility (Petrides & Vila de Brito, 2024). Petrides and Vila de Brito (2024) further contend that while digital platforms do not guarantee sales for artists, they have benefits such as assisting with acquiring followers and building professional connections where job prospects may present themselves. This assertion is evident in the practicing artist's career, where the use of social media in 2019 significantly contributed to her gaining a huge following on social media and an opportunity to exhibit her work in the USA. Mäkelä (2024) also posits that when there is a slight feeling of fame around an artist on social media, it creates a sense of credibility which can result in people engaging with the artist's work.

Both alumni significantly increased their professional visibility by leveraging the social media. Their experiences reveal the distinction between proactive digital engagement and reactive viral impact. The viral post gave the practicing artist a huge following and an opportunity to showcase their work in the USA. In contrast, the fashion designer actively launched a new collection virtually during the 2020 lockdown, strategically using social media to adapt to challenging circumstances. However, while social media boosts visibility, it can also commodify art. To mitigate this limitation, alumni prioritize artistic integrity, meaning they refuse to alter their authentic voices to gain popularity. Instead, they emphasize consistency and personal branding.

7.4.6 Consistency Long-Term Success, Perseverance and Commitment

Consistency is an essential element in an artist's path to success, as it influences audience recognition of their work. Kore (2023) explained that consistency means behaving or performing similarly. In art, this is understood as an ongoing, recognizable style or artistic voice. For instance, a practicing artist's signature style involves the use of synthetic hair as a medium of exploration (Radebe, 2019). In an interview with Ownurcrown (2019) the practicing artist emphasized the importance of maintaining a specific artistic style that viewers can immediately recognize and how having a distinctive artistic approach contributes to the success of the artist. A practicing artist stated:

"Consistency is absolutely essential for any artist to be successful. As an emerging artist, I must remain focused and not confuse myself or the viewers. I want to make it easier for viewers to identify the work. I want to send a coherent message to my viewers and not send mixed messages; this way, people will understand my work and hopefully be inspired" (Ownurcrown 2019).

On the other hand, a fashion designer known for creating high-end fashion designs for over 20 years has emphasized that consistency is an important aspect that an artist needs to build a successful brand (Mazibuko, 2023). In an interview with Bella Naija Style (2019), when the fashion designer was asked how they knew their brand was truly headed for success. The fashion designer expressed that: "With time and consistency I started reaping the results" (Bella Naija Style, 2019).

Both alumni view consistency in their work as a vital factor for success in their field. The practicing artist believes that consistency assists the audience in recognizing and understanding the meaning of their work, which contributes to the artist's success. In contrast, the fashion designer attributed consistency to leading positive outcomes over time.

7.5 Conclusions and Recommendations

This chapter aims to measure student outcomes by documenting the success stories of the alumni from the VAD department at the UoT in South Africa. Through an examination of the career trajectories and achievements of the three alumni—a fashion designer, an independent curator, and a practicing artist—five factors emerged as critical for alumni success: professional mentorship, collaboration with major brands, competitive achievement, use of social media, and consistency and branding. First, professional mentorship played a significant role in assisting alumni to learn more about the business of art post-university. Second, partnerships with big-name artists and established brands contributed to alumni's professional development and success by enhancing their exposure, opportunities, and new target markets. Third, participation and winning 1st prize in competitions have contributed to the alumni's success by paving the way for launching a successful brand, providing grand exposure, and further exhibition opportunities. Fourth, leveraging social media for augmented influence and professional visibility to promote and showcase art has proven to be a success factor for the alumni. Finally, consistency which includes having a unique artistic signature, has been identified as an important aspect of long-term success and recognition.

While the UoT program provided the three alumni with technical skills that enabled them to start a successful career, their success and achievements were advanced by extraneous engagements, such as professional mentorship, collaboration, competitions, social media, and consistency. The identified need for mentoring foregrounds a crucial gap that needs to be addressed to ensure that the UoT's mandate for professional preparedness is achieved. Conversely, factors such as collaboration with big brands, winning competitions, strategic use of social media, and consistency and branding reinforce UoT's commitment to professional preparedness and industry relevance. These elements equip alumni with real-world experience, boosting their visibility, digital presence, and recognition in the creative industry.

The success factors identified in this study imply a compelling need for UoTs to transcend conventional ways of measuring student success through academic performance and completion rate to alumni outcomes as measures to evaluate student success. Therefore, these findings align with those of Spence (2019), Blankstein and Wolff-Eisenberg (2020), Salazar and Schelbe (2021), and Mathews (2023), who argued that students' success should be measured through alumni. Such an approach enables graduates to point out how the university program has influenced their successful careers and what is still lacking in the university program to ensure that the UoT produces industry-ready graduates.

Furthermore, the findings imply that the art industry has an equal obligation to nurture talent in the field. This includes actively engaging with higher institutions through mentorships and partnerships to provide real-world opportunities. Moreover, to effectively align the curriculum with the ever-changing industry demands, UoTs should formally integrate alumni expertise into their policies. Through alumni advisory boards, UoTs can receive firsthand information on crucial curriculum developments.

Therefore, this study contributes to the growing body of research advocating for the use of alumni as measures of student success in higher education.

As part of the continuous improvement process of teaching and learning at the UoT, this study calls for a revision of the VAD curriculum to ensure that it produces graduates who will have long-term sustainable art careers. To improve the curriculum, this study recommends the following:

Currently, the VAD department at the UoT, where the three alumni studied, lacks formalized mentorship programs across all four of its sections (fine art, photography, fashion design, and graphic design). This study recommends that the institution formalize mentoring through alumni in the VAD department to assist students in transitioning into professional careers. Mentorship assists in career satisfaction (DeCastro et al., 2014), which in turn improves retention and experience (Bonifacino et al., 2021).

The correct use of social media can significantly enhance teaching and learning for VAD students. Therefore, this study recommends that the curriculum should integrate content that leverages social media as a pedagogical tool. Ajumobi and Mahomed (2024) assert that students need to develop digital competency to thrive amidst technological advancement in higher education and prepare for sustainable success. The integration of social media comes with benefits such as marketing and promotion of students' work, increased visibility in the online space, brand building through consistent digital presence, engagement with the audience through direct messages, the creation of opportunities for collaboration and partnership, driving sales, and acquiring new clients. Furthermore, social media can be used for alumni interaction with current students. The use of technology aligns with the mission of the UoT, as articulated by Chinomona et al. (2013).

To further enhance the VAD curriculum at UoT, a structured module on artistic consistency and branding should be integrated into the curriculum. This can be achieved through workshops focused on developing artistic identity, where students can cultivate their unique styles. Additionally, lessons on personal branding and audience engagement should be included to build students' confidence and identity. These sessions could be facilitated by successful artists currently active in the industry or accomplished VAD Alumni.

Currently, within the four sections of the VAD department, there are annual competitions that students are encouraged to participate in. For example, in fashion, third-year students are encouraged to participate in the annual Bernina competition. Fine Arts students are encouraged to enter annual competitions such as the Thami Mnyele Awards, Sasol Signature, and 24-h animation contest for students, among others. However, despite this encouragement, only a small proportion of students participated. This study recommends integrating competitions into the curriculum through competition-based assessments. This approach will ensure that all students participate in competitions, thereby familiarizing them with the custom of creating and presenting their art for external assessment. Lecturers can provide essential guidance on the application process, proposal writing, and portfolio compilation.

The study recommends that university-industry partnerships be explicitly integrated into the mission and vision of the UoT. This formal integration ensures that such collaborations are included in the curriculum. Dowelani and Dowelani (2020) note that university-industry partnerships are essential for skills development, as they ensure that skills taught in higher education align with industry requirements. Hence, through sustained collaboration and ongoing communication, academic institutions and businesses can mutually inform each other about the necessary curricular modifications. The UoT would significantly benefit from this by becoming more flexible in creating in-demand programs that are pertinent to the changing demands of businesses (Matthew, 2023). Moreover, the engagement of successful alumni with the VAD department can also assist in approaching prominent brand partners.

In line with the qualitative approach, a major limitation of this study is its small sample size of three participants. Therefore, the findings cannot be generalized to the entire population of VAD graduates from UoTs in South Africa. To improve generalization, further studies should aim for a larger and more diverse sample of VAD alumni from different UoTs in South Africa.

Although secondary data are convenient and cost-effective, they have inherent limitations. These include instances such as the research objective of the original data not aligning with the current study, a lack of detailed information, and limited control over quality data points. In addition, it obviates the opportunity to ask follow-up questions to participants to gain an in-depth understanding of their success factors. Therefore, further studies should prioritize primary data collection by directly engaging UoT VAD alumni through interviews, focus groups, and surveys. This approach would allow for rich data that are specifically tailored to answer the study's research question.

Another limitation is the potential bias in the media representation of alumni. The media frequently prioritizes notable achievements, which can lead to bias and an incomplete representation of alumni career paths. This optimistic view might disregard the failures and setbacks experienced by alumni. Thus, further studies should prioritize self-reported data to address media biases. This method would allow alumni to share their complete career narratives, challenges, and small achievements that media data sources often overlook.

Future studies could conduct longitudinal studies to track alumni career trajectories over time. This approach would allow for the validation of this study's findings and provide unbiased and detailed accounts of alumni career paths over an extended period.

References

Ajumobi, D. O., & Mahomed, L. (2024). COVID-19's impact on students' digital identity and academic success at South African Higher Education Institutions. In *Redefining Education and Development: Innovative Approaches in the Era of the Sustainable Development Goals* (pp. 113–131). Springer Nature Switzerland. https://doi.org/10.1007/978-3-031-69954-2

Arora, S. (2022). Fugitive aesthetics: Performing refusal in four acts. In *Intimacy and injury* (pp. 309–336). Manchester University Press. https://works.hcommons.org/records/rz6h2-k2a82

Artist Proof Studio. (2025). *Lebohang Motaung*. https://artistproofstudio.co.za/pages/lebohang#:~:text=Lebohang%20Motaung's%20work%20has%20been%20showcased%20at,artwork%20has%20garnered%20within%20the%20art%20community. Accessed 12 Feb 2025.

Balotescu, C. (2024). *How an alumni engagement program boosts student career readiness*. https://www.360alumni.com/blog/alumni-engagement-program-boosts-student-career-success. Accessed: 30 Mar 2025.

Bandura, A. (1977). Self-efficacy: Toward a unifying theory of behavioral change. *Psychology Review, 84*(2), 191–215. https://doi.org/10.1037/0033-295X.84.2.191

Barron, J. (2015). Building a chain of success in marketing higher education: The alumni connection. *Industrial and Commercial Training, 47*(5), 253–256. https://doi.org/10.1108/ICT-01-2015-0005

Basri, M., & Sitorus, A. S. (2023). The role of alumni in curriculum development in early childhood Islamic education program (PIAUD) UIN Sumatra Utara Medan. *Edukasi Islami: Jurnal Pendidikan Islam, 12*(01): 353–366. https://doi.org/10.30868/ei.v12i01.3492

Bella Naija Style. (2019). *Catching up with David Tlale—The fashion king of South Africa spills his secrets to success*. https://www.bellanaijastyle.com/exclusive-david-tlale-bn-style-secrets/ Accessed 12 Apr 2025.
Bieger, T. (2024). The Role of alumni in the development of their University. In *Business School Internationalisation in a Changing World* (pp. 23–27). Routledge. https://doi.org/10.4324/9781003500629-4
Blankstein, M., & Wolff-Eisenberg, C. (2020). Measuring the whole student: landscape review of traditional and holistic approaches to community college student success. *Ithaka S+ R*. https://doi.org/10.18665/sr.313888
Bonifacino, E., Ufomata, E. O., Farkas, A. H., Turner, R., & Corbelli, J. A. (2021). Mentorship of underrepresented physicians and trainees in academic medicine: A systematic review. *Journal of General Internal Medicine, 36*, 1023–1034. https://doi.org/10.1007/s11606-020-06478-7
Brand South Africa. (2013). *Brand SA congratulates David Tlale*. https://brandsouthafrica.com/102724/empowerment/brand-sa-congratulates-david-tlale/. Accessed 07 Mar 2025.
Brand South Africa. (2025). *South African designers take to the world stage at Shanghai fashion week*. https://brandsouthafrica.com/177911/community/south-african-designers-take-to-the-world-stage-at-shanghai-fashion-week/. Accessed 05 Apr 2025.
Braun, V., & Clarke, V. (2006). Using thematic analysis in psychology. *Qualitative Research in Psychology, 3*, 77–101. https://doi.org/10.1191/1478088706qp063oa
Byrne, D. (2022). A worked example of Braun and Clarke's approach to reflexive thematic analysis. *Quality & Quantity, 56*(3), 1391–1412. https://doi.org/10.1007/s11135-021-01182-y
Brynjolson, N. (2017). *From mentorship to collaboration: Art, feminism, and community in Winnipeg*. Contemporary Feminist Art in Canada, p. 220. http://www.jstor.org/stable/j.ctt1q31s5m.16
Chinomona, E., Chinomona, R., & Moloi, K. C. (2013). Elements of quality assurance at institutions of higher education: Vaal University of Technology in South Africa. *Mediterranean Journal of Social Sciences, 4*(14), 643–656. https://doi.org/10.5901/mjss.2013.v4n14p643
Creswell, J. W. (2014). *Research design: Qualitative, quantitative, and mixed methods approaches*. Sage Publications.
Creswell, J. W., & Poth, C. N. (2016). *Qualitative inquiry and research design: Choosing among five approaches*. Sage Publications.
Dawson, R. F., Kearney, M. S., & Sullivan, J. X. (2020). *Comprehensive approaches to increasing student completion in higher education: A survey of the landscape* (No. w28046). National Bureau of Economic Research.
Dean, K. W., & Jendzurski, M. B. (2021). Achieving student success through alumni re-activation: A virtual interpersonal engagement approach. *Journal of Access, Retention, and Inclusion in Higher Education, 4*(1), 86–96. https://digitalcommons.wcupa.edu/cgi/viewcontent.cgi?article=1028&context=jarihe
DeCastro R, Griffith K. A., Ubel.P. A., Stewart A., & Jagsi R. (2014). Mentoring and the career satisfaction of male and female academic medical faculty. *Acad Med, 89*(2):301–311. https://doi.org/10.1097/ACM.0000000000000109
DiBenedetto, M. K., & Schunk, D. H. (2022). Assessing academic self-efficacy. In M. S. Khine & T. Nielsen (Eds.), *Academic self-efficacy in education: Nature, assessment, and research* (pp. 11–37). Springer Nature Singapore Pte Ltd. https://doi.org/10.1007/978-981-16-8240-7_2
Digital Drum. (2024). *Lebohang Motaung on her career and showcasing at the Joburg Art Fair*. https://www.snl24.com/drum/celebs/news/lebohang-motaung-on-her-career-and-showcasing-at-the-joburg-art-fair-20240906. Accessed 16 Mar 2025.
Dowelani, M., & Dowelani, F. (2020). *Curriculum Development in South Africa: The Role of Professional Bodies*. https://doi.org/10.4995/head20.2020.11188
Dzimbili, S. (2019). *Local artist on her hair artwork that went viral*. https://www.snl24.com/drum/hair/local-artist-on-her-hair-artwork-that-went-viral-20190719. Accessed 08 Mar 2025.

Enihe, R., & Omopariola, V. (2023). Alumni portal system for nigerian universities (a case study). *International Journal of Science and Research, 12*(2), 1047–1055. https://doi.org/10.21275/SR23217134355

Forbes Africa. (2019). *Fashion, fame and finances with SA designer, David Tlale.* https://www.forbesafrica.com/7-questions-with/2018/08/03/fashion-fame-and-finances-with-sa-designer-david-tlale/. Accessed 16 Mar 2025.

Glamour South Africa. (2024). *Discover the vibrant world of printmaking through the Artist Proof Studio and its up-and-coming talent at FNB Art Joburg.* https://www.glamour.co.za/lifestyle/discover-the-vibrant-world-of-printmaking-through-the-artist-proof-studio-and-its-up-and-coming-talent-at-fnb-art-joburg-93f3dc2e-2163-480b-8cfe-87997197d162. Accessed 12 Aug 2024.

Graham, L., Williams, L., & Chisoro, C. (2019). Barriers to the labour market for unemployed graduates in South Africa. *Journal of Education and Work, 32*(4), 360–376. https://doi.org/10.1080/13639080.2019.1620924

Guest, G., MacQueen, K. M., & Namey, E. E. (2012). *Applied thematic analysis.* Sage publications.

Herrera-Limones, R., Rey-Pérez, J., Hernández-Valencia, M., & Roa-Fernández, J. (2020). Student competitions as a learning method with a sustainable focus in higher education: The University of Seville "Aura Projects" in the "Solar Decathlon 2019." *Sustainability, 12*(4), 1634. https://doi.org/10.3390/su12041634

Ibáñez-Sánchez, S., Flavián, M., Casaló, L. V. & Belanche, D. (2022). Influencers and brands successful collaborations: A mutual reinforcement to promote products and services on social media. *Journal of Marketing Communications, 28*(5), 469–486. https://doi.org/10.1080/13527266.2021.1929410

Independent Curator International. (2013). *Nkululeko Khumalo.* https://curatorsintl.org/about/collaborators/5372-nkululeko-khumalo. Accessed 26 Sep 2024.

Jennings, H. (2015). A brief history of postcolonial African fashion. *Fashion and Postcolonial Critique, 22*, 104–113. https://doi.org/10.1215/10757163-3339860

Keteyi, O. (2022). *David Tlale, brand magnet: A look at how he chooses the brands he collaborates with.* https://iol.co.za/lifestyle/style-beauty/fashion/2022-05-14-david-tlale-the-brand-magnet-a-look-at-how-he-chooses-the-brands-he-collaborates-with/. Accessed 17 Mar 2025.

Kempton Park Express. (2018). *Watch: SA's best artists honoured in Kempton Park.* https://www.citizen.co.za/kempton-express/news-headlines/2018/10/04/sas-best-artists-honoured-in-kempton-park. Accessed 10 Feb 2025.

Kool, A., Mainhard, M. T., Jaarsma, A. D. C., van Beukelen, P., & Brekelmans, M. (2018). Do students with varying academic ability benefit equally from personal qualities? Applying a trait and state perspective. *Research in Higher Education, 59*, 1021–1034. https://doi.org/10.1007/s11162-018-9498-y

Kore. (2023). Consistency in art making. https://koresageart.com/blog/consistency-in-art-making/. Accessed 13 Mar 2025

Kuh, G. D., Kinzie, J. L., Buckley, J. A., Bridges, B. K., & Hayek, J. C. (2006). What matters to student success: A review of the literature (Vol. 8). National Postsecondary Education Cooperative. https://nces.ed.gov/npec/pdf/kuh_team_report.pdf

Kwarteng, H. O., & Obeng-Ofori, D. (2021). Enhancing the role of alumni in the growth of higher education institutions. *International Journal of Multidisciplinary Studies and Innovative Research, 4,* 40–48. https://scholar.google.com/scholar?q=Obeng-Ofori%2C+D.%2C+%26+Kwarteng%2C+H.+O.+%282020%29.+Enhancing+the+role+of+alumni+in+the+growth+of+higher+education+institutions.+International+Journal+of+Multidisciplinary+Studies+and+Innovative+Research%2C+3%283%29%2C+40%E2%80%9348

Latitudes. (2024). *Lebohang Motaung.* https://latitudes.online/artists/lebohang_motaung. Accessed 01 Apr 2025.

Matovu, M. M. (2020). Academic self-efficacy and academic performance among university undergraduate students: An antecedent to academic success. *European Journal of Education Studies, 7*(3), 135–149. https://doi.org/10.5281/zenodo.3756004

Mäkelä, O. (2024). *The role of social media content exposure, perceived artist authenticity, and enduring involvement in consumer attitudes toward artists.* https://helda.helsinki.fi/server/api/core/bitstreams/12bb6d67-23b7-4f8b-9eae-1f51511c3f8f/content. Accessed 13 Mar 2025.

Matthew, A. (2023). Alumni as a benchmark of student success. https://www.timeshighereducation.com/campus/alumni-benchmark-student-success. Accessed 26 Sep 2024.

Mazibuko, T. (2023). *From design to mentoring and education: Celebrating 20 years of fashion with David Tlale.* https://www.iol.co.za/sunday-tribune/lifestyle/from-designs-to-mentoring-and-education-celebrating-20-years-of-fashion-with-david-tlale-8aec06bd-0828-431a-8b87-3c5a8e7a1453. Accessed 03 Mar 2025.

McCollum, G. (2018). *Collegiate recreation participation and student retention, progression, and graduation.* Doctoral dissertation. Georgia Southern University. https://digitalcommons.georgiasouthern.edu/cgi/viewcontent.cgi?article=2993&context=etd

Memani, Q. (2019). *VUT established art curator making waves offshore.* https://vut.ac.za/vut-established-art-curator-making-waves-offshore/. Accessed 26 Aug 2024.

Mukhari, A. (2021). *David Tlale: An Icon living.* https://www.glamour.co.za/fashion/david-tlale-an-icon-living-297865df-41af-4b4d-aaa2-8f5f6a95d3f3. Accessed 04 Apr 2025.

Nanda, H. I., Putri, S. F., Putri, D. M., Ermayda, R. Z., & Palil, M. R. (2021). Study of alumni engagement and its relationship to University Curriculum reforming. In *ICE-BEES 2020: Proceedings of the 3rd International Conference on Economics, Business and Economic Education Science, ICE-BEES 2020*, 22–23 July 2020, Semarang, Indonesia (p. 354). European Alliance for Innovation. https://doi.org/10.4108/eai.22-7-2020.2307898

Ncube, T. R., & Lekhanya, L. M. (2025). The role of entrepreneurship education in shaping final-year students' career paths: A case study of Durban University of Technology's faculty of management science. *Journal of Economic and Social Development (JESD)–Resilient Society, 12*(1). https://www.proquest.com/openview/607c5c015781924ce6c3f1b4a9233981/1?pq-origsite=gscholar&cbl=2037666

Nkosi, M. (2021). *Veteran lensman Mike Mzileni says cameras are more powerful than guns at new exhibition launch.* City Press. https://www.news24.com/citypress/trending/veteran-lensman-mike-mzileni-says-cameras-are-more-powerful-than-guns-at-new-exhibition-launch-20211213. Accessed 14 June 2026.

Nony, A. (2019). Racism and Culture in the age of techno-aesthetic supremacy. *The polish journal of aesthetics, 52*(1), 13–20. https://pjaesthetics.uj.edu.pl/documents/138618288/142419802/pja-52-preface-nony.pdf/adac7b8e-4223-43ab-a40e-889f1a854beb

Odero, C. (2020). *World acclaimed South African designer on the business of fashion.* https://nation.africa/kenya/life-and-style/lifestyle/world-acclaimed-south-african-designer-on-the-business-of-fashion-1125684. Accessed 16 Mar 2025.

Oranga, J., & Matere, A. (2023). Qualitative research: Essence, types and advantages. *Open Access Library Journal, 10*(12), 1–9. https://doi.org/10.4236/oalib.1111001

Ownurcrown. (2019). *Lebohang Motaung—Hair artist & stylist.* https://ownurcrown.business/2017/07/19/hair-art/. Accessed 26 Sep 2024.

Papadopoulos, T., Taylor, T., Fallshaw, E., & Zanko, M. (2011). *Engaging industry: Embedding professional learning in the business curriculum.* Australian Learning and Teaching Council. https://researchportal.murdoch.edu.au/esploro/outputs/report/Engaging-industry-Embedding-professional-learning-in/991005540484707891/filesAndLinks?index=0

Peters, W. (2021). DUT advancement and alumni relations office launches DUT alumni mentorship programme. https://www.dut.ac.za/dut-advancement-and-alumni-relations-office-launches-dut-alumni-mentorship-programme/#:~:text=Being%20a%20University%20that%20wishes,%2C%20Alumni%20Relations%20Co%2Dordinator Accessed 10 June 2025.

Petrides, L., & Fernandes, A. (2020). The successful visual artist: The building blocks of artistic careers model. *The Journal of Arts Management, Law, and Society, 50*(6), 305–318. https://doi.org/10.1080/10632921.2020.1845892

Petrides, L., & Vila de Brito, M. (2024). The impact of digital presence on the careers of emerging visual artists. *Social Sciences, 13*(6), 313. https://doi.org/10.3390/socsci13060313

Pico-Saltos, R., Garzás, J., Redchuk, A., Escandón-Panchana, P., & Morante-Carballo, F. (2022). Role of alumni program in the prediction of career success in an Ecuadorian Public University. *Applied Sciences, 12*(19), 9892. https://doi.org/10.3390/app12199892

Radebe, Z. L. (2019). *Exploring black women's diverse hairstyles through art: a case study*. Master's dissertation. Vaal University of Technology. https://digiresearch.vut.ac.za/items/850f8299-a20d-4ba2-b0ac-eff1cf0e3a7c

Riebe, L., & Jackson, D. (2014). The use of rubrics in benchmarking and assessing employability skills. *Journal of Management Education, 38*(3), 319–344. https://doi.org/10.1177/1052562913511437

Saunders-Smits, G., & de Graaff, E. (2012). Assessment of curriculum quality through alumni research. *European Journal of Engineering Education, 37*(2), 133–142. https://doi.org/10.1080/03043797.2012.665847

Schneider, M., & Preckel, F. (2017). Variables associated with achievement in higher education: A systematic review of meta-analyses. *Psychological Bulletin, 143*(6), 565–600. https://doi.org/10.1037/bul0000098

Salazar, A. M., & Schelbe, L. (2021). Factors associated with post-college success for foster care alumni college graduates. *Children and Youth Services Review, 126*, Article 106031. https://doi.org/10.1016/j.childyouth.2021.106031

Soika, B. (2021). *What is student success?* New insight into a complex question. https://rossier.usc.edu/news-insights/news/what-student-success-new-insight-complex-question. Accessed 16 Mar 2025.

Spence, C. (2019). Judgement' versus 'metrics' in higher education management. *Higher Education, 77*(5), 761–775. https://doi.org/10.1007/s10734-018-0300-z

Steinmayr, R., Meiñer, A., Weideinger, A. F., & Wirthwein, L. (2014). *Academic achievement* (pp. 9780199756810–0108). Oxford University Press.

Triangle Network Projects. (2013). Triangle project: Curatorial intensive hosted by Bag Factory and ICI, Johannesburg. https://www.trianglenetwork.org/triangle-network/projects/triangle-project-curatorial-intensive-hosted-by-bag-factory-and-ici-johannesburg-7-12-march-2013. Accessed 14 June 2026.

Tulankar, S., & Grampurohit, B. (2020). Role of alumni as stakeholders in enhancing quality education. In *Sustaining quality: NAAC new guidelines 2017 conference, Bhandup, Mumbai*. https://www.researchgate.net/publication/341650267_Role_of_alumni_as_stakeholders_in_enhancing_quality_education

Woodside, J. M. (2018). Real-world rigour: An integrative learning approach for industry and higher education. *Industry and Higher Education, 32*(5), 285–289. https://doi.org/10.1177/0950422218784535

Yende, S. J. (2021). Performing arts: A case study on curriculum transformation. *The Independent Journal of Teaching and Learning, 16*(1), 129–141

Yende, S. J., & Mugovhani, G. (2021). Employability challenges facing vocal art graduates in South Africa: A case study of Tshwane University of Technology. *Muziki, 18*(1), 110–126. https://doi.org/10.1080/18125980.2021.1957002

Yiba. (2021). *Bringing photography back to its roots while honouring a legend*. https://yiba.co.za/bringing-photography-back-to-its-roots-while-honouring-a-legend/. Accessed 26 Feb 2025.

York, T. T., Gibson, C., & Rankin, S. (2015). Defining and measuring academic success. *Practical Assessment, Research & Evaluation, 20*(5), 5. https://doi.org/10.7275/hz5x-tx03

Zhuang, T., & Tao, Z. (2024). What lessons can university education learn from outside the ivory tower: Insights from engineering alumni. *Teaching in Higher Education, 29*(4), 970–986. https://doi.org/10.1080/13562517.2022.2048367

Part IV
Digital Transformation and Technological Infrastructure

Chapter 8
Information and Communication Technology Education Level II: Subjects Delivered Through Technology-Enhanced Teaching and Learning

Johannes Segone Dirane

Abstract The integration of technology into Information and Communications Technology (ICT) education has significantly transformed traditional teaching practices, especially in ICT Education Level II subjects. This study examines the effectiveness of three technology-enhanced learning (TEL) tools, Learning Management Systems (LMS), Gmetrix, and Jasperactive, in improving student performance at a South African University of Technology. Quantitative data from 1920 participants collected between 2021 and 2023 were analysed to evaluate the impact of these tools on pass rates, student engagement, and digital competency. The results show substantial improvements in pass rates, which rose from 76.52% to 86.30% in 2021, from 65.94% to 88.51% in 2022, and from 64.20% to 85.50% in 2023. Notably, online assessments consistently outperformed the face-to-face assessments. The findings indicate that TEL promotes digital literacy, enhances higher-order thinking skills, encourages self-directed learning, and better prepares students for digital workplaces. The study recommends that institutions strategically invest in educator training and adopt blended learning models to maximise the benefits of TEL. By providing continuous assessments, personalised feedback, and flexible access to learning materials, these platforms help bridge educational gaps and offer a scalable approach to improving academic outcomes in ICT education in the future.

Keywords Technology-enhanced learning (TEL) · ICT Education Level II subjects · E-learning · Gmetrix · Learning Management System (LMS) · Jasperactive

J. S. Dirane (✉)
Centre for Academic Development, Vaal University of Technology, Vanderbijlpark, South Africa
e-mail: johannesd2@vut.ac.za

K. Dube and L. M. Tjabane (eds.), *Advancing Sustainable Development Goals in Higher Education*, https://doi.org/10.1007/978-3-032-27646-9_8

8.1 Introduction and Background

The Sustainable Development Goals (SDGs), established by the United Nations, recognise the transformative role of education and technology in building inclusive, equitable, and sustainable societies. SDG 4, which focuses on Quality Education, aims to "ensure inclusive and equitable quality education and promote lifelong learning opportunities for all." This goal emphasises the integration of Information and Communication Technology (ICT) to enhance teaching and learning, support digital literacy, and address educational inequalities (Arora & Mishra, 2019). It promotes the use of technology to reach marginalised populations, improve learning outcomes, and provide access to open educational resources and e-learning platforms (Eden et al., 2024).

Equally important is SDG 9, which focuses on Industry, Innovation, and Infrastructure. This goal underscores the significance of fostering innovation and expanding access to information and communication technologies, particularly in underdeveloped regions (Sinha et al., 2024). It supports the creation of resilient digital infrastructure that empowers educational systems, enables remote learning, and nurtures future-ready skills such as coding, data analysis, and problem-solving. By aligning educational efforts with digital transformation, countries can ensure that learners are equipped to thrive in knowledge-based economies, reduce the digital divide, and contribute meaningfully to sustainable development (Upadhyaya, 2024).

Technology integration in teaching and learning has fundamentally transformed the global educational landscape (Rahimi & Oh, 2024). Technology enhances accessibility and facilitates personalised learning experiences, establishing itself as an indispensable tool in contemporary education (Kalyani, 2024). Research from various international contexts has identified several critical areas in which technology has a significant influence (Basilotta-Gómez-Pablos et al., 2022; Choi-Lundberg et al., 2023; Stannard & Sallı, 2019; Wang et al., 2023). Technology is a vital educational resource that provides essential materials and tools that facilitate learning (Kalyani, 2024). This includes digital online books, eBooks, virtual courses, computer-based learning, and computer-based training software. Educational applications present students with various interactive and engaging learning resources. Furthermore, technology supports the implementation of remote and hybrid learning models, which have gained prominence, particularly in the aftermath of the COVID-19 pandemic (Carretero Gomez et al., 2021).

Educational technology is crucial for developing the essential skills needed to use technology effectively in teaching and learning environments. (Kalyani, 2024). This includes Learning Management Systems such as Moodle and Blackboard, as well as video conferencing tools such as Zoom and Microsoft Teams. Platforms for coding, simulation software, and virtual laboratories enable students to acquire and practice their competencies in a controlled environment (Yalova et al., 2025). These technological innovations also upgrade the planning and management processes within the educational system, thereby facilitating the effective organisation and administration of educational accomplishments (Alenezi, 2023). Extensive global research underscores the transformative potential of technology within the realm of education, including enhanced accessibility of inclusion, personalised learning, interactive and engaging learning, improved educator support and professional development, data-driven insights, and global collaboration (Gowher Alam & Mohanty, 2023b; Hassan, 2023; Kumar et al., 2023). By strategically harnessing technological advancements, educators and curriculum developers can improve inclusive, engaging, and effective learning environments that address the unique needs of students worldwide (Eden et al., 2024).

Technology integration in teaching and learning has emerged as an essential element of educational advancement in South Africa (Ohei et al., 2023; Shambare & Simuja, 2024). As the nation attempts to close educational disparities and enhance learning outcomes, technology is critical for redefining traditional educational methods (Oke & Fernandes, 2020). It is an essential tool for addressing historical imbalances in education. Initiatives such as the Digital Doorway project illustrate the capacity of technology to facilitate access to computer literacy skills in underserved communities (Heimann et al., 2025; van der Vyver & Marais, 2015). These initiatives have demonstrated that children can develop basic computer skills through self-directed learning, emphasising the significance of peer collaboration. The application of Information and Communication Technologies (ICT) in South African schools and universities has progressed from basic computer-aided instruction to more advanced methodologies, encompassing mobile learning and social media integration (Bonsu et al., 2020). These developments have permitted the creation of personalised learning experiences and enhanced student engagement. Nevertheless, technology adoption in education is confronted with challenges, including economic disparities, infrastructural limitations, and varying levels of digital literacy, which may impede the effective implementation of teaching and learning using technology-enhanced learning (Jacobs & Pretorius, 2020). To address these challenges, it is imperative to adopt comprehensive strategies that encompass training for educators, regulatory governance, and provision of equitable access to technology. National research emphasises the transformative potential of technology in South Africa's educational landscape (Ajani, 2024). By utilising advancements in technology-enhanced learning, educators can create more inclusive, engaging, and effective environments that cater to students' diverse needs across the country.

Teaching and learning Information and Communication Technology (ICT) education Level II subjects involve two key components: theory and practical application. The theory component covers the fundamental concepts of hardware and operating software, including their functions, differences, and applications, which are the learning outcomes of the subjects (Andersson, 2023). The practical component emphasises hands-on experience with application software (Microsoft Word, Excel, and PowerPoint) across introductory, intermediate, and advanced levels, which aligns with the subjects' learning outcomes (Nouwens & Nylandsted Klokmose, 2021). Students are trained in using essential programmes such as Microsoft Word, Excel and PowerPoint, along with operating software in the computer setup. They learn to navigate and use software applications and manage and process information effectively using a personal computer (PC). The Level II ICT Education curriculum fosters both conceptual understanding and practical skills, equipping students with real-world ICT applications.

The traditional ICT Level II subject syllabus heavily relied on face-to-face instruction, which limited flexibility and self-paced learning (Abad & Abad, 2023). The COVID-19 pandemic further exposed the need for technology-driven learning solutions (Adedoyin & Soykan, 2023) This study investigates how the integration of LMS, Gmetrix, and Jasperactive has influenced student engagement and learning outcomes in ICT Education Level II subjects. In those years, students were supposed to attend practical classes physically in their different computer laboratories on campus (Dirane, 2007). Every educator taught the students specific concepts during every lesson, for example, how to use the mail merge function in Microsoft Word, IF statement(s) in Microsoft Excel, or how to set up a proper presentation in Microsoft PowerPoint. Once the concept was presented to the students, the educator transferred the working file (an unfinished version) to the students' personal computers. The students then followed the educator step-by-step to organise the working file into a completed product (Nouwens & Nylandsted Klokmose, 2021).

This study is based on the Technological Pedagogical Content Knowledge (TPACK) framework developed by Koehler et al. (2014). This framework offers a comprehensive model for understanding the knowledge educators need to effectively integrate technology into their teaching. TPACK extends Shulman's (1986) concept of Pedagogical Content Knowledge (PCK) by adding Technological Knowledge (TK) as a third essential domain (Shing et al., 2018). The framework emphasises the intersection of three core areas: Content Knowledge (CK), Pedagogical Knowledge (PK), and Technological Knowledge (TK). Effective technology integration occurs when educators have a solid understanding of the subject matter (CK), how to teach it (PK), and how to utilise technology (TK) in ways that enhance learning. In this study, TPACK is used as an analytical lens to examine how pre-service educators design digital learning activities, highlighting the interplay between their technological fluency, pedagogical strategies, and content expertise.

In 2020, the nationwide lockdown began on 26 March due to COVID-19, preventing physical attendance in computer laboratories on campus. The COVID-19 pandemic forced a sudden and widespread transition from traditional face-to-face instruction to technology-enhanced learning (TEL) at all levels of education (Smallman & Ryan, 2023). While this shift was necessary to maintain continuity of learning during lockdowns and social distancing measures, it revealed significant gaps in digital readiness among educators, students, and institutions (Kundu & Bej, 2021). Many educators who had previously relied on in-person teaching methods were unprepared to design and deliver effective online or blended instruction. As students could not physically attend classes, the possibilities for learning became virtually limitless. University educators need to devise solutions to continue teaching and learning effectively. According to McHaney (2023), students today are technologically knowledgeable (tech-savvy) because they can access many resources and information at their fingertips. Subsequently, the educators at the university had to develop a new strategy for teaching and learning with students using technology, which involved using the Learning Management System (LMS), Gmetrix, and Jasperactive software to teach students online theory and online practical components (Dirane, 2024).

A Learning Management System is a digital learning environment that manages all the features of a learning institution or a company's various training efforts (Koh & Kan, 2021). An LMS is a centralised platform for content delivery, communication, and performance tracking (Bradley, 2021). Gmetrix is a certification preparation resource often used for Microsoft Office Specialist (MOS) certifications, designed to help students and professionals prepare for certification by providing practice exams (Richardson, 2023). Gmetrix provides certification exam preparation through forums and group assignments (White et al., 2020). Jasperactive is a cutting-edge learning system hosted on Microsoft Azure that teaches students various fundamental skills in technology, such as the basics of using computers and the Internet, business accounting, and Microsoft Office, delivered through the Microsoft Office suite of applications (Mokonyane et al., 2021). The purpose of Jasperactive is to provide learners with a tool applicable to all kinaesthetic learning styles that can be accessed anywhere and provides the freedom to learn at any pace (Dirane & Ismail, 2024).

The emergence of digital learning platforms and the widespread adoption of Learning Management Systems (LMS), Gmetrix, and Jasperactive have redefined ICT education. These technologies provide interactive, self-paced, and immersive learning experiences, allowing students to develop both theoretical knowledge and practical skills remotely (Rutkauskaite et al., 2022). The shift towards technology-enhanced learning became even more pronounced during the COVID-19 pandemic, when lockdown measures necessitated alternative learning strategies (Carretero Gomez et al., 2021). The abrupt transition from physical classrooms to virtual platforms underscored the importance of integrating digital tools into ICT education to ensure continuity and accessibility (Alenezi et al., 2023; Best & MacGregor, 2017).

While technology-driven instruction brings many advantages, some significant hurdles remain, such as gaps in digital literacy, unequal access to technology resources, and questions about the effectiveness of online teaching, learning, and assessments (Ogange et al., 2020). This study aligns with the United Nations' South Africa goal of promoting inclusive and equitable quality education. It explicitly targets Goal 4 and its sub-target 4.1, which focuses on lifelong learning opportunities for students (Sever et al., 2025). Integrating technology-enhanced learning (TEL) into South African education directly supports Sustainable Development Goal 4 (SDG 4) by increasing equitable access to quality learning resources and promoting lifelong learning. Technology-enhanced learning addresses systemic barriers such as geographic isolation, financial constraints, and limited infrastructure, particularly in underserved communities, through mobile learning, open educational resources, and digital skills training (Ibrahim & Jibia, 2024).

8.2 Literature Review

Technology revolutionises education by transforming how, when, and where students learn, empowering students and educators on their educational journeys (Dirane & Ismail, 2024). This allows students to learn online and collaborate by sharing ideas and experiences (Brandauer et al., 2022). With limitless access to a wealth of information, technology knows no bounds. Mobile devices, laptops, tablets, smartphones, and computers have become an integral part of daily life, enabling students and educators to tap into endless digital learning resources (Alarifi, 2023; Dirane, 2024; Matthew et al., 2018). Moreover, embracing technology in teaching and learning allows students to personalise their educational experiences. It empowers them to take charge of how they grasp information, making education significant and impactful for digital natives committed to lifelong learning in the twenty-first century. Using 16th-century teaching methods for 21st-century students is impractical due to significant technological advancements (Dirane, 2016).

Teaching and learning technology tools, such as virtual reality (VR), video conferencing platforms, online books, eBooks, and various educational technologies, can make traditional classrooms and virtual learning environments more engaging (Childs et al., 2024). Educators are expected to acquire advanced knowledge and diverse strategies to use technology effectively to enhance the learning experience, support instruction, and facilitate student learning (Haleem et al., 2022). Integrating technology into teaching can foster a more inclusive classroom where students are encouraged to collaborate and explore (Fayzullayev et al., 2024). With access to relevant online content, learning applications, real-time student data, and more, technology can empower students to reach new heights in their educational careers (McHaney, 2023). Additionally, educators can create mixed or blended learning platforms and environments, leverage digital tools for formative and summative assessments, and introduce innovative approaches to teaching and learning in traditional classrooms and settings for visually impaired students (Adera, 2024).

Shatrı (2020) outlined several advantages of integrating information technology into classroom settings. One of the most significant benefits is the increase in student motivation and interest in learning, as digital tools often make the content more engaging and interactive. Technology also allows for individualised instruction, enabling educators to address diverse learning needs and learning paces. The use of visually dynamic and rapidly changing presentations stimulates cognitive engagement and supports the development of creative skills.

Moreover, technology promotes objectivity in assessments and helps students build their information literacy skills. AI empowers students to make quick, informed decisions in complex situations and provides immediate access to extensive information databases, which streamlines research and enhances learning efficiency (Kalyani, 2024). The use of digital platforms has also been linked to a higher rate of assignment completion and encouraged independent study habits (Khanduri & Teotia, 2023).

Additionally, a variety of technology-enhanced activities, including educational games, promote deeper mental engagement (Childs et al., 2024). Online education, in particular, broadens the flow of information and offers new opportunities for students to participate more actively in the learning process, leading to improved educational outcomes. Shatrı (2020) argued that using information technology has the following benefits: it fosters greater collaboration, offers limitless resources, aids in developing future-ready skills, conserves valuable resources, updates information instantly, and takes up little space.

In recent years, integrating technology-enhanced learning platforms has transformed educational practices across various disciplines by reshaping how content is delivered, accessed, and engaged with. Technology platforms encourage interactive and participatory learning. STEM fields use simulations and virtual laboratories, such as PhET and MATLAB Simulink, to allow students to experiment in a safe and cost-effective environment (Javali et al. 2025). Humanities and Social Sciences employ discussion forums, blogs, and multimedia tools to foster debate, crucial analyses, and peer collaboration (Suleimanova et al., 2020). Professionals in fields such as health sciences can upskill through asynchronous courses while working full time (Geshwaree & Harish, 2024). In language learning, applications such as Duolingo use adaptive algorithms to reinforce weak areas, improving retention and motivation (Tuong & Dan, 2024). In disciplines such as business or education, dashboards track participation, quiz scores, and engagement to inform time intervention. AR/VR in medicine and architecture for immersive learning (Thangavel, 2025). AI in computer sciences and finance to teach advanced analytics, automation, and ethical considerations (Abdussalam & Auladi, 2024). Multimedia assignments (videos, podcasts, digital storytelling) are especially used in arts and education, promoting creativity and communication (Parekh, 2025).

Numerous studies have highlighted the benefits of TEL in improving students' engagement, accessibility, and personalised learning experiences (Alotaibi, 2024; Daniela et al., 2018; Dunn & Kennedy, 2019; FitzGerald et al., 2018; Major et al., 2021; Sajja et al., 2025). However, much of the existing research has concentrated on the general education of Science, Technology, Engineering, and Mathematics (STEM) subjects, with limited focus on ICT Education Level II subjects, particularly in using LMS, Jasperactive, and Gmetrix platforms. Furthermore, while some studies have explored the effectiveness of TEL tools, few have provided empirical evidence on how these platforms specifically influence students' learning outcomes in ICT-focused educator training or professional development contexts. This gap is particularly significant, given the increasing dependence on digital tools in educational environments. Therefore, this study addresses this gap by examining how the TEL platform impacts student performance and learning experiences in ICT Education Level II subjects. The study also evaluated how technology-enhanced learning platforms could improve student outcomes within ICT Education Level II subjects.

This study focuses on the role of Information Communication Technology (ICT) in Level II subjects delivered through technology-driven teaching and learning. This is significant because it investigates how digital tools and platforms can improve the quality, accessibility, and effectiveness of advanced education. As students advance to more specialised and complex subjects, the integration of ICT enhances understanding, fosters critical thinking, and promotes essential 21st-century skills, such as digital literacy, problem-solving, and independent learning. Additionally, technology facilitates more inclusive and flexible learning environments that accommodate diverse student needs and learning styles.

Furthermore, this study contributes to the achievement of Sustainable Development Goals, particularly SDG 4 (Quality Education) and SDG 9 (Industry, Innovation, and Infrastructure) by encouraging innovative teaching practices and preparing students for a digital, knowledge-based economy. This research provides valuable insights that can help guide institutions and policymakers in making informed decisions about ICT integration in higher education.

8.3 Materials and Methods

This study adopted a quantitative approach to evaluate the effectiveness of technology-enhanced learning in improving student outcomes within the ICT Education Level II curriculum offered by the Department of Education at a selected University of Technology. Participants enrolled in the ICT Education Level II subjects were engaged with online teaching and learning platforms, specifically LMS, Gmetrix, and Jasperactive. These technologies provide interactive, self-paced, and immersive learning experiences that enable students to develop both theoretical knowledge and practical skills, both face-to-face and remotely. After participating in these learning experiences, students from all groups took the tests, and the researcher recorded all assessment scores, both online and practical, on the Integrated Technology System (ITS) as part of the instruction for the ICT Education Level II subjects.

To ensure an organised and secure assessment process, detailed arrangements were made to conduct each computer-based test in the university's computer laboratories. Before the test date, students were informed about the schedule, laboratory allocations, and access credentials through the institution's Learning Management System. Each student was required to arrive at least 30 min before the start of each test for verification and seating arrangements. Upon arrival, students presented their identification cards and were assigned to specific workstations according to preplanned seating charts. This was done to minimise distractions and reduce the risk of academic misconduct. The tests were administered through a secure online platform under the strict supervision of invigilators and technical support staff. All non-permitted electronic devices and materials were prohibited inside the laboratory.

If any technical issues arose during the session, on-site technicians were available to promptly address and resolve them to ensure minimal disruption. After completing each test, students had to submit their responses digitally through the LMS, Gmetrix, and Jasperactive platforms and exit the laboratory quietly, allowing a conducive environment for those who were still taking the test. These structured arrangements ensured fairness, security, and consistency in the administration of computer-based assessments.

At the end of each academic year, the researcher used the ITS computing system to analyse the assessment scores. The system generates results using the computing elements in the ITS for each student on the ITS list across all groups. It then calculates the average scores, standard deviations, grade types, and overall marks for both practical and online testing formats for each group separately. These data sources were selected for their reliability and relevance to the objectives of this study.

A quantitative methodology is justified in this study because of its capacity to yield measurable and objective data that can be analysed statistically (Mohajan, 2020). The collected data were quantitative and consisted of numerical measurements. This includes scores from the digital competency assessment in ICT Education Level II subjects, covering all the test scores for each student. The data included average scores, standard deviations, grade types, and overall marks, all of which were measurable, countable, and suitable for statistical analysis. This approach facilitates the systematic exploration of the relationship between technology-enhanced learning and student performance, allowing for the formulation of clear conclusions based on empirical evidence to be made. By implementing established software tools and a comprehensive data collection framework, this study aims to ensure the accuracy and dependability of its findings.

This study investigated the impact of technology-enhanced learning (TEL) platforms on student outcomes in ICT Education Level II subjects. The primary research question aimed to determine how Learning Management Systems (LMS), Gmetrix, and Jasperactive affect pass rates and student engagement in ICT Education Level II subjects at the South African University of Technology (UoT). The research objectives include assessing the impact of TEL platforms on student performance, comparing outcomes between students using TEL platforms and those employing traditional learning methods, and identifying the specific features of TEL platforms that contribute most significantly to improved outcomes. By analysing student performance data from 2021 to 2023, this study offers valuable insights into the role of digital tools in enhancing academic achievement, student engagement, and practical skill acquisition. The findings contribute to the ongoing discussion on the adoption of educational technology and provide recommendations for optimising ICT education in higher education institutions.

8.3.1 Population

The research population consisted of participants registered in the ICT Education Level II subjects within the Department of Education at the selected University of Technology during 2021, 2022, and 2023. This cohort represented a complete set of individuals possessing the necessary characteristics for the study. At the end of each specified year, a census was conducted to gather data from all groups whose computed marks were recorded in the Integrated Technology Services (ITS) list for ICT Education Level II subjects. This population was selected based on the need to understand the educational outcomes associated with technology teaching, learning, and assessment at the selected University of Technology.

By focusing on this demographic, the researcher aims to provide insights relevant to the broader context of technology integration in teaching and learning, which may lead to improved educational outcomes. Participants were intentionally selected based on their direct experience with the integration of digital platforms, tools, and methodologies in the teaching and learning processes. By concentrating on this targeted group, this study sought to capture relevant data on how technology impacts student engagement, understanding, and performance in Level II ICT education subjects. This sampling method ensured that the participants had the necessary background and exposure to provide informed and contextually relevant feedback aligned with the study's objectives. Figure 8.1 visually represents the total number of students registered yearly.

In 2021, 496 students registered for ICT Education Level II subjects. This figure increased to 582 in 2022. The participants were organised based on the Integrated Technology Services (ITS) list and their respective class schedules, which were designated as groups A, B, C, D, E, F, G, and H for the period from January to December in both years. In 2023, the number of students enrolled in ICT Education II rose to 842. This year, the participants were also categorised according to the ITS list, with their class timetable categorised as group A2, B2, C2, DA, DB, DC, DD, DE, DF, E3, F2, PA, PB, and XA, from January to December in 2023. The total number of participants through those years is 1920.

This research entails a comprehensive census of all participants enrolled in the ICT Education II subject within the Department of Education at the selected University of Technology. This methodology ensures that the data collected are thorough and accurately reflect the entire population (Gilmore et al., 2022). The census procedures

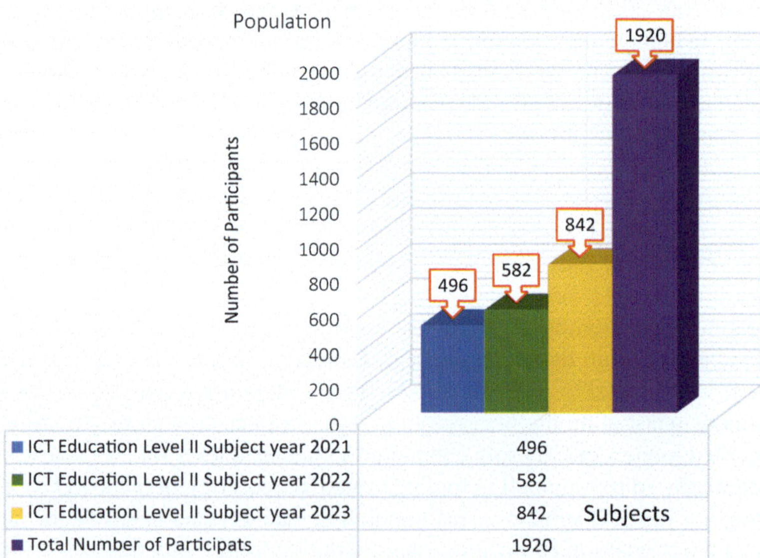

Fig. 8.1 Population of ICT education level II students in 2021, 2022, and 2023

involved the collection of records from the computerised ITS list for all face-to-face and online theory assessments, as well as for all practical assessments conducted in both formats for all groups. Students' engagement followed these procedures in face-to-face and online or virtual teaching and learning environments. This approach aims to provide a holistic understanding of the characteristics and experiences of the population under study.

8.3.2 Data Collection

This study combined primary and secondary data collection methods to obtain comprehensive and reliable results. Primary data were continuously obtained from the Integrated Teaching System (ITS) class lists for all groups registered for ICT Education subjects, with online and practical assessments recorded in 2021, 2022, and 2023. The ICT Education Level II syllabus consists of theoretical and practical components. The theoretical component addresses fundamental topics such as hardware, software, and their functions and purposes. Conversely, the practical component emphasises software applications at the basic, intermediate, and partially advanced levels, explicitly focusing on Microsoft Word, Excel, and PowerPoint. During 2021 and 2022, the syllabus was delivered through online and face-to-face methods. However, in 2023, it transitioned to a fully online format, incorporating tools such as LMS, Gmetrix, and Jasperactive software. These technologies provide interactive, self-paced, and immersive learning experiences, allowing students to develop both theoretical knowledge and practical skills physically and remotely (Rutkauskaite et al., 2022).

The assessments for ICT Education Level II subjects included tests 1, 2, 3, 4, and 5. Each time participants completed a test, whether online or in a practical format, their scores were recorded within the ITS platform. This platform, including ITS Integrator and ITS Infinity by Adapt IT, functions as a centralised system that connects various operational areas (Sandberg et al., 2020). It integrates student admissions, registrations, residence management, bursary administration, assessments, examinations, graduation processing, and alumni relations into a single cohesive environment. Subsequently, after each academic year, the marks entered were computed according to the designated percentage weight for each test. This process yielded reliable calculated marks and comprehensive scores for each student and facilitated the determination of average scores, standard deviations, mark type marks, and full period marks for each group. Using the ITS records, the study aimed to analyse the calculated average scores and standard deviations to ascertain whether technology integration positively or negatively influenced pass rates. This methodological approach provided a robust framework for undertaking a detailed analysis of the research objective, which sought to evaluate the extent to which technology-enhanced learning platforms can improve student outcomes within ICT Education Level II subjects.

8.3.3 *Data Analysis*

The data analysis process employed quantitative techniques to comprehensively examine the collected data. Quantitative data were gathered from the computed Integrated Technology System (ITS) records for all groups registered in the ICT Education Level II subjects during 2021, 2022, and 2023. The system produces results using the computing elements that educators operate towards the end of November or during the first week of December every year. This process occurs after all student test data have been corrected and captured in the ITS system for each student on the ITS list across all groups. It then calculates the average scores, standard deviations, grade types, and overall marks for both practical and online testing formats for each group separately.

The quantitative analysis focused on numerical and statistical evaluations, employing measures such as frequency, mean, standard deviation, and central tendency for each sample within the population under study (Bruce et al., 2018). The ICT Education Level II subjects encompassed five assessments: tests 1, 2, 3, 4, and 5. Continuous data analysis was conducted on written and online theoretical tests and practical assessments completed by students between January and December 2021, 2022, and 2023. In 2021, online test 3 was conducted to assess the content delivered through the LMS, Gmetrix, and Jasperactive software platforms. In 2022, online tests 1 and 3 were evaluated similarly based on the content taught using these online tools.

By 2023, all five online assessments, tests 1, 2, 3, 4, and 5 were conducted to evaluate the material presented through LMS, Gmetrix, and Jasperactive software. The results of the data analysis are presented through graphs generated from spreadsheets using Microsoft Excel software and discussion. This analytical methodology facilitated a thorough understanding of the research findings, allowing nuanced interpretations of the computed ITS list results. Throughout the analysis process, meticulous attention was devoted to maintaining the integrity and reliability of the data, ensuring that the conclusions drawn were well supported and valid.

8.4 Results and Discussions

The findings of this study yield significant insights into the differences in averages and standard deviations between practical and online testing formats. The quantitative data revealed both negative and positive discrepancies in the outcomes of face-to-face versus online teaching and learning approaches, suggesting that implementing online teaching and learning methodologies may improve the pass rate for ICT Education Level II subjects. Figure 8.2 visually represents the comparative averages and standard deviations for face-to-face and online students and the percentage of participants who registered for the ICT Education II subject across all groups and academic years.

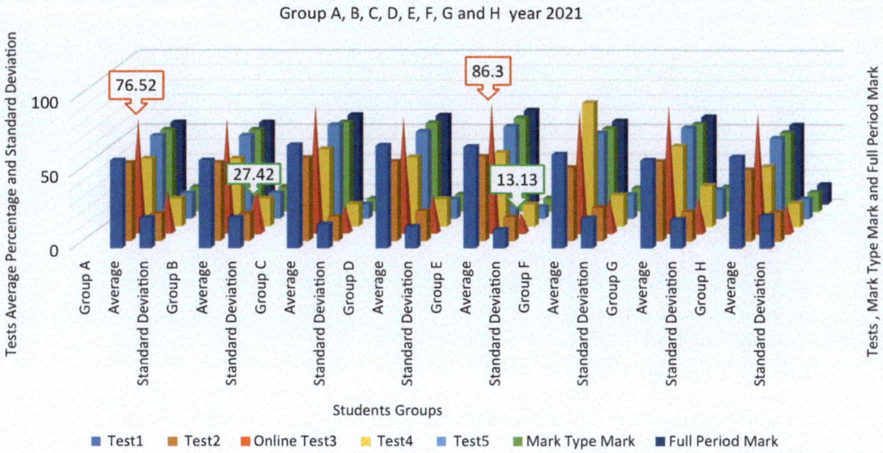

Fig. 8.2 Tests' averages and standard deviations for Groups A, B, C, D, E, F, G and H in the year 2021

The study provides significant insights into the differences in average scores, standard deviations, mark types, and overall marks between practical and online testing formats for groups A, B, C, D, E, F, G, and H in 2021. Students performed significantly better in online assessments (Test 3), with average scores ranging from 76.52% to 86.3% and standard deviations between 13.13% and 27.42%, as shown in Fig. 8.2. In contrast, face-to-face tests (Tests 1, 2, 4, and 5) had average scores between 40.11% and 82.72%, with standard deviations ranging from 8.2% to 22.63%. The findings indicate that the integration of technology in teaching and learning has the potential to positively impact academic performance when employed judiciously. Future research should investigate strategies to assist students in harmonising their face-to-face and online educational experiences with their academic obligations. Furthermore, implementing interventions that encourage the mindful use of online learning modalities may prove advantageous in alleviating the negative consequences on academic performance. Figure 8.3 visually represents the comparative averages and standard deviations for face-to-face and online students and the percentage of participants who registered for the ICT Education II subject across all groups and academic years.

The findings of this study provide significant insights into the differences in average scores, standard deviations, mark types, and overall marks between practical and online testing formats for groups A, B, C, D, E, F, G, and H in 2022. Students performed significantly better in online assessments (Test 1), with average scores ranging from 65.94% to 73% and standard deviations between 13.92% and 25.83% in the two tests. Students performed significantly better in online assessments (Test 3), with average scores ranging from 76.38% to 88.51% and standard deviations between 6.56% and 24.48%. In contrast, face-to-face tests (Tests 1, 2, 4, and 5) had average scores between 40.48% and 58.77% and standard deviations ranging from

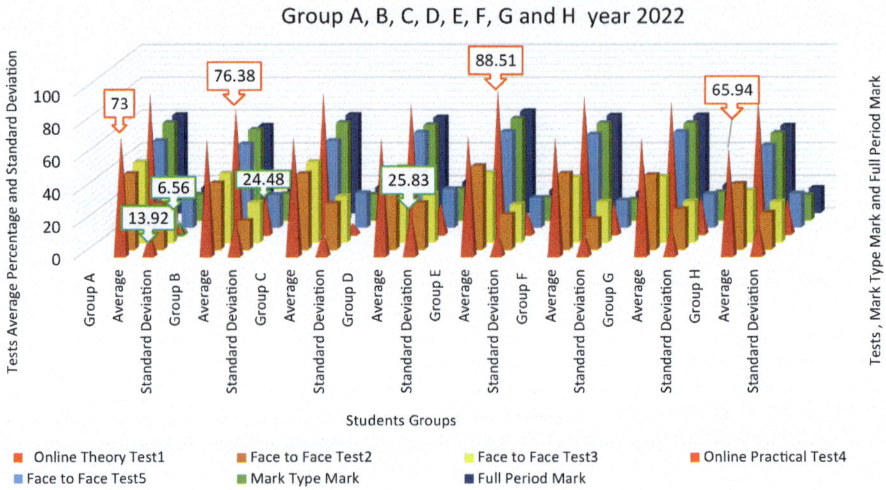

Fig. 8.3 Tests' averages and standard deviations for Groups A, B, C, D, E, F, G and H in the year 2022

16.88% to 22.68%, as demonstrated in Fig. 8.4. The findings indicate that integrating technology into teaching and learning can positively impact academic performance when employed cautiously. Future research should investigate strategies to assist students in harmonising their face-to-face and online educational experiences with their academic obligations. Furthermore, implementing interventions that encourage the mindful use of online learning modalities may prove advantageous in alleviating the negative consequences on academic performance.

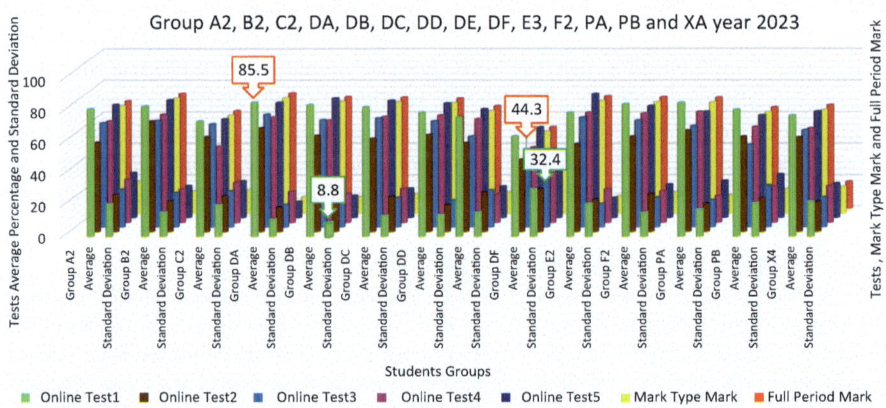

Fig. 8.4 Test averages and standard deviations for Groups A2, B2, C2, DA, DB, DC, DD, DE, DF, E3, F2, PA, PB, and XA in 2023

The findings of this study provide significant insights into the differences in average scores, standard deviations, mark types, and overall marks between practical and online testing formats for Groups A2, B2, C2, DA, DB, DC, DD, DE, DF, E3, F2, PA, PB, and XA in 2023. Students performed significantly better in all online assessments (Tests 1, 2, 3, 4, and 5), with average scores ranging from 44.3% to 85.5% and standard deviations between 8.8% and 32.4%, as demonstrated in Fig. 8.4. The findings indicate that integrating technology into teaching and learning can positively affect academic performance when employed judiciously. Future research should identify effective strategies to support students in consistently using online educational experiences in conjunction with their academic obligations. Furthermore, implementing interventions that promote the intentional use of online learning modalities may be beneficial in moderating adverse effects on academic performance.

This study investigated how technology-enhanced learning platforms can improve student outcomes in ICT Education Level II subjects. The findings were systematically categorised into three areas: student engagement with LMS, Gmetrix, and Jasperactive; learning outcomes; and skill development. Participants engaged with LMS, Gmetrix, and Jasperactive, platforms that employ online learning through interactive theoretical content and practical exercises centred on computer fundamentals and the application of Microsoft Word, Excel, and PowerPoint. These exercises were designed to be self-paced, offering immersive learning experiences that facilitated the development of theoretical knowledge and practical skills in face-to-face and remote settings.

Participants could access the online software and content from diverse locations, allowing them to progress at their own pace and convenience. The use of online tools supports the development of essential skills such as problem-solving and collaboration and correlates with improved learning outcomes, as test scores indicate. This aligns with Kalyani (2024), who posited that technology is a crucial educational resource that provides essential materials and tools that enhance the learning experience. Furthermore, it is consistent with the observations made by Carretero Gomez et al. (2021), who highlighted that technology facilitates the implementation of remote and hybrid learning models, particularly in the context of the ongoing adjustments following the COVID-19 pandemic.

The results of this study revealed significant positive effects attributable to the use of technology on participant engagement and academic performance in higher education contexts. Quantitative data analysis revealed that participants who regularly used educational technology platforms such as LMS, Gmetrix, and Jasperactive, which were specifically designed for the ICT Education Level II curriculum, demonstrated higher levels of engagement. These platforms offer interactive, self-paced, and immersive learning experiences, enabling students to develop both theoretical knowledge and practical skills in physical classrooms or through remote learning. As a result, this engagement leads to better academic performance. Specifically, their average scores ranged from 44.3% to 88.51%, compared to average scores of 31.8% to 82.72% for those who did not engage with these technological resources.

The integration of technology rendered the learning process more interactive and accessible, thereby enhancing the participants' comprehension of complex concepts and practical applications. This aligns with Fayzullayev et al. (2024), who support the integration of technology in teaching to create a more inclusive classroom environment and promote student collaboration and exploration. These findings underscore the potential of integrating technology into educational curricula to significantly enhance effective student engagement and learning outcomes.

This finding lines up directly with Alam and Mohanty (2023a), Gowher Hassan (2023), and Kumar et al. (2023), who stated that the extensive global research underscores the transformative potential of technology with the national realm of education embracing, among other things, enhanced accessibility of inclusion, personalised learning, interactive and engaging learning, improved educator support and professional development, data-driven insights, and global, national, and local collaboration. This also directly aligns with Shatrı's (2020) findings, which stated that information technology has the following benefits: it fosters greater collaboration, offers limitless resources, aids in developing future-ready skills, conserves valuable resources, updates information instantly, and takes up little space. They highlight the necessity of increased investment in educational technology resources to facilitate effective teaching and learning.

8.5 Recommendations

Based on the analysis of Learning Management Systems, Gmetrix, and Jasperactive in the context of ICT Education Level II subjects at South African Universities of Technology, it is recommended that educational institutions increase their investment in a wide range of technology-enhanced learning tools. These platforms have been effective in improving student pass rates, engagement, and digital literacy through personalised feedback, interactive content, and skills development aligned with industry requirements. To build on this success, institutions should incorporate additional tools such as virtual laboratories, interactive simulations, virtual reality (VR), video conferencing platforms, digital textbooks, and eBooks. These technologies cater to diverse learning styles and provide immersive experiences that help students understand complex theoretical and practical concepts in the field. A strategic and inclusive approach to integrating technology will not only enhance academic performance but also better prepare students for the demands of the modern workforce.

To maximise the impact of technology-enhanced teaching and learning, educational institutions must prioritise the establishment of comprehensive training programmes for educators. These programmes should not only focus on building technical proficiency but also on equipping educators with pedagogical strategies that facilitate the seamless integration of technology into various teaching methods to enhance learning. This enables educators to effectively utilise digital platforms to create dynamic, engaging, and student-centred environments. Furthermore, institutions must ensure that continuous technical support is provided and access to appropriate resources for both students and educators.

This ongoing support is crucial in addressing any technical difficulties that may arise during the learning process, thereby promoting a smoother and more inclusive educational experience for all students. Additionally, institutions should adopt blended learning models that combine traditional classroom instruction with online and flexible learning environments. Such models offer greater flexibility, allowing students to learn at their own pace and according to their individual needs while still benefiting from structured face-to-face interactions. This hybrid approach fosters personalised learning experiences and accommodates diverse learning styles, ultimately enhancing academic outcomes and student satisfaction.

To fully leverage the potential of technology in education, institutions should establish continuous evaluation and feedback mechanisms to assess the effectiveness of the technologies used. This approach allows for informed adjustments and keeps the educational environment engaging, inclusive, and effective. Institutions should prioritise enhancing accessibility and inclusion, support personalised learning experiences, promote interactive and engaging content delivery, and strengthen professional development for educators. Additionally, using data-driven insights and fostering collaboration at the global, national, and local levels can significantly improve teaching and learning practices. By adopting these strategies, educational institutions can ensure that technology is a transformative tool that enhances both the quality and equity of education.

The University of Technology (UoT) should organise the ICT Education Level II curriculum into weekly modules, with clearly aligned Intended Learning Outcomes (ILOs), Teaching and Learning Activities (TLAs), and Assessment Tasks (ATs). To enhance student engagement, it is important to include multimedia content such as videos, quizzes, and discussion forums. Monitoring students' progress using Learning Management System (LMS) analytics can also help identify and support at-risk students. Encouraging peer collaboration through forums and group assignments further enriches the learning experience.

It is recommended that Gmetrix practice tests be integrated into the Learning Management System (LMS) at the University of Technology as weekly assignments. Additionally, pre- and post-assessments should be used to measure the skill development and readiness. Targeted feedback based on Gmetrix performance should be provided to guide students' revision. Furthermore, Gmetrix modules should be aligned with Jasperactive tasks to reinforce learning. Suggestions for utilising Jasperactive at the University of Technology during computer laboratory sessions aim to enhance theoretical concepts through practical tasks include: These activities should allow students to progress independently, supporting differentiated learning. Additionally, self-directed learning can be encouraged by assigning Jasperactive modules as homework. Using Jasperactive's, real-time feedback helps students correct their mistakes and build confidence.

8.6 Conclusion

The use of technology in teaching and learning is highly recommended because of its positive impact on student performance. The results indicate that the average pass rate for participants in all groups increased significantly to above 44.3% and reached more than 88.51% in 2023. In contrast, the test scores from 2021 and 2022 showed a lower average of 31.80%. However, this figure improved in subsequent assessments, with test scores ranging between 40.11% and 88.51% across different tests and student groups in the same year. This trend demonstrates that educators in future classrooms should embrace modern technology. Using the latest tools, they can offer students unlimited access to essential theoretical and practical knowledge, tailored to their needs. Therefore, it is crucial for educational institutions, including educators and curriculum developers, to recognise the importance of integrating technology into their teaching practices.

Embracing 21st-century tools is essential for helping digital natives quickly understand information, making education relevant and impactful in this century, and encouraging lifelong learning, which aligns with the United Nations' Sustainable Development Goal 4 (SDG 4). Proven statistics from research analyses highlight that technology can significantly enhance teaching and learning outcomes, particularly in subjects related to ICT Education II.

The implementation of various teaching and learning methodologies has significantly improved digital literacy, increased student engagement, fostered higher-order thinking skills, prepared students for the digital workplace, and encouraged self-directed learning and autonomy in the learning process. Technology-enhanced learning platforms have transformed education by making it more accessible, interactive, personalised, and data-informed. Each discipline experiences distinct advantages, but all share a common shift towards student-centred, flexible, and collaborative educational ecosystems.

References

Abad, D. J. V., & Abad, E. M. (2023). Self-Paced learning practices during educational disruption in a public higher education institution. *East Asian Journal of Multidisciplinary Research, 2*(5), 1957–1978. https://doi.org/10.55927/eajmr.v2i5.3819

Abdussalam, A. V., & Auladi, G. A. (2024). Pushing boundaries: AI and computer science in the era of technological revolution. *TechComp Innovations: Journal of Computer Science and Technology, 1*(1), Article 1. https://doi.org/10.70063/techcompinnovations.v1i1.22

Adedoyin, O. B., & Soykan, E. (2023). Covid-19 pandemic and online learning: The challenges and opportunities. *Interactive Learning Environments, 31*(2), 863–875. https://doi.org/10.1080/10494820.2020.1813180

Adera, N. (2024). Innovative learning spaces and blended learning: Quest for 21st century competency teaching and learning approaches. In G. N. Areba, B. N. Gisore, & E. W. Njurai (Eds.), *Advances in educational marketing, administration, and leadership* (pp. 139–174). IGI Global. https://doi.org/10.4018/979-8-3693-6930-2.ch006

Ajani, O. A. (2024). Sustainable transformation in South African Rural Universities: A digital perspective. *International Journal of Management, Knowledge and Learning, 13*(1), 113–127. https://www.ceeol.com/search/article-detail?id=1255123

Alam, A., & Mohanty, A. (2023a). Developing 'happiness engineering' subject for the schools in India: Designing the pedagogical framework for a sustainable happiness curriculum. *Qubahan Academic Journal, 3*(4), 1–20.

Alam, A., & Mohanty, A. (2023b). Educational technology: Exploring the convergence of technology and pedagogy through mobility, interactivity, AI, and learning tools. *Cogent Engineering, 10*(2), 2283282. https://doi.org/10.1080/23311916.2023.2283282

Alarifi, I. M. (2023). *Learning limitless knowledge: Transformative learning pathway to unlocking and harnessing the endless power of knowledge.* Balboa Press.

Alenezi, M. (2023). Digital learning and digital institution in higher education. *Education Sciences, 13*(1), Article 1. https://doi.org/10.3390/educsci13010088

Alenezi, M., Wardat, S., & Akour, M. (2023). The need of integrating digital education in higher education: Challenges and opportunities. *Sustainability, 15*(6), Article 6. https://doi.org/10.3390/su15064782

Alotaibi, N. S. (2024). The impact of AI and LMS integration on the future of higher education: opportunities, challenges, and strategies for transformation. *Sustainability, 16*(23), Article 23. https://doi.org/10.3390/su162310357

Andersson, J. C. (2023). *Learning microsoft azure.* O'Reilly Media, Inc.

Arora, N. K., & Mishra, I. (2019). United nations sustainable development goals 2030 and environmental sustainability: Race against time. *Environmental Sustainability, 2*(4), 339–342. https://doi.org/10.1007/s42398-019-00092-y

Basilotta-Gómez-Pablos, V., Matarranz, M., Casado-Aranda, L.-A., & Otto, A. (2022). Teachers' digital competencies in higher education: A systematic literature review. *International Journal of Educational Technology in Higher Education, 19*(1), 8. https://doi.org/10.1186/s41239-021-00312-8

Best, M., & MacGregor, D. (2017). Transitioning design and technology education from physical classrooms to virtual spaces: Implications for pre-service teacher education. *International Journal of Technology and Design Education, 27*(2), 201–213. https://doi.org/10.1007/s10798-015-9350-z

bipach. (n.d.). *Microsoft office specialist: Expert (Office 2019)—Certifications.* Retrieved March 19, 2025, from https://learn.microsoft.com/en-us/credentials/certifications/microsoft-office-specialist-expert-2019/

Bonsu, N. O., Bervell, B., & Arkorful, V. (2020). Computer-Assisted instruction in the teaching and learning of history: A systematic review in Africa. *International Journal of Innovation, 14*(9).

Bradley, V. M. (2021). Learning management system (LMS) use with online instruction. *International Journal of Technology in Education, 4*(1), 68–92. https://eric.ed.gov/?id=EJ1286531

Brandauer, S., Carnine, J., DeGuzman, K., Grazioli, B., Lyons, L., Sandiford, N., & Hartman, E. (2022). Crisis as opportunity: Reimagining global learning pathways through new virtual collaborations and open access during COVID-19. *Frontiers: The Interdisciplinary Journal of Study Abroad, 34*(1), 9–23. https://eric.ed.gov/?id=EJ1346247

Bruce, N., Pope, D., & Stanistreet, D. (2018). *Quantitative methods for health research: A practical interactive guide to epidemiology and statistics.* John Wiley & Sons.

Carretero Gomez, S., Napierala, J., Bessios, A., Mägi, E., Pugacewicz, A., Ranieri, M., Triquet, K., Lombaerts, K., Robledo Bottcher, N., Montanari, M., & Gonzalez Vazquez, I. (2021). What did we learn from schooling practices during the COVID-19 lockdown. *LUX*. https://doi.org/10.2760/135208

Childs, E., Mohammad, F., Stevens, L., Burbelo, H., Awoke, A., Rewkowski, N., & Manocha, D. (2024). An overview of enhancing distance learning through emerging augmented and virtual reality technologies. *IEEE Transactions on Visualization and Computer Graphics, 30*(8), 4480–4496. https://doi.org/10.1109/TVCG.2023.3264577

Choi-Lundberg, D. L., Butler-Henderson, K., Harman, K., & Crawford, J. (2023). A systematic review of digital innovations in technology-enhanced learning designs in higher education. *Australasian Journal of Educational Technology, 39*(3), Article 3. https://doi.org/10.14742/ajet.7615

Daniela, L., Visvizi, A., Gutiérrez-Braojos, C., & Lytras, M. D. (2018). Sustainable higher education and technology-enhanced learning (TEL). *Sustainability, 10*(11), 3883. https://doi.org/10.3390/su10113883

Dirane, J. S. (2007). Teaching and learning entrepreneurship online in Soshanguve Technical High School: A case study. In Montgomerie, C. & Seale, J. (Eds.), *Proceedings of ED-MEDIA 2007—World conference on educational multimedia, hypermedia & telecommunications* (pp. 2251–2258). Vancouver, Canada: Association for the Advancement of Computing in Education. (AACE). Retrieved January 18, 2023, from https://www.learntechlib.org/primary/p/25684/

Dirane, J. S. (2016). *The state of using mobile devices in ubiquitous learning environments at two selected universities of technology*. PhD Thesis, University of Technology, Pretoria.

Dirane, J. S. (2024). Teaching and learning using technology. In *The future of television and video industry*. IntechOpen. https://doi.org/10.5772/intechopen.114246

Dirane, J. S., & Ismail, Y. (2024). Teaching and learning using technology. In *The future of television and video industry*. IntechOpen. https://doi.org/10.5772/intechopen.114246

Dunn, T. J., & Kennedy, M. (2019). Technology enhanced learning in higher education; motivations, engagement and academic achievement. *Computers & Education, 137*, 104–113.

Eden, C., Onyebuchi, N., & Adeniyi, I. (2024). Harnessing technology integration in education: Strategies for enhancing learning outcomes and equity. *World Journal of Advanced Engineering Technology and Sciences, 11*, 001–008. https://doi.org/10.30574/wjaets.2024.11.2.0071

Fayzullayev, K., Esanova, N., & Bozorov, H. (2024). "Collaborative classrooms: building inclusive learning environments." *Multidisciplinary Journal of Science and Technology, 4*(12), Article 12. http://www.mjstjournal.com/index.php/mjst/article/view/2191

FitzGerald, E., Kucirkova, N., Jones, A., Cross, S., Ferguson, R., Herodotou, C., Hillaire, G., & Scanlon, E. (2018). Dimensions of personalisation in technology-enhanced learning: A framework and implications for design. *British Journal of Educational Technology, 49*(1), 165–181. https://doi.org/10.1111/bjet.12534

Geshwaree, H., & Harish, C. (2024). *Revitalizing student skills for workforce preparation*. IGI Global.

Gilmore, G., Randich, S., Worley, C. C., Hourihane, A., Gonneau, A., Sacco, G. G., ... & Fiorentin, P. R. (2022). The Gaia-ESO public spectroscopic survey: Motivation, implementation, GIRAFFE data processing, analysis, and final data products. *Astronomy & Astrophysics, 666*, A120.

Gowher Hassan. (2023). Technology and the transformation of educational practices: A future perspective. *International Journal of Economic, Business, Accounting, Agriculture Management and Sharia Administration (IJEBAS), 3*(1), 1596–1603. https://doi.org/10.54443/ijebas.v3i1.1136

Haleem, A., Javaid, M., Qadri, M. A., & Suman, R. (2022). Understanding the role of digital technologies in education: A review. *Sustainable Operations and Computers, 3*, 275–285. https://doi.org/10.1016/j.susoc.2022.05.004

Heimann, C., Minnie, S., & Abrams, M. C. (2025). UMGCINI—Empowering rural communities through geospatial data and digital connectivity. *The International Archives of the Photogrammetry, Remote Sensing and Spatial Information Sciences, XLVIII-M-5–2024*, 59–64. ASPRS 2024 Annual Conference—at Geo Week, 11–13 February 2024, Denver, Colorado, USA and 21–24 October 2024 (virtual conference). https://doi.org/10.5194/isprs-archives-XLVIII-M-5-2024-59-2025

Ibrahim, T. A., & Jibia, S. A. (2024). Utilizing technology for education: Exploring innovative ways in education to bridge the digital divide in underserved communities. *International Journal of Library Science and Educational Research*. https://cambridgeresearchpub.com/ijlser/article/view/278

Jacobs, J., & Pretorius, M. W. (2020). The major challenges facing organisations to create technology-enabled value in the fourth industrial revolution: A dynamic capabilities perspective in South Africa. *South African Journal of Industrial Engineering, 31*(2), 40–61. https://doi.org/10.7166/31-2-2345

Javali, P, Raj, R. G., Shaw, S., Devi, C. H., Kumar, A., Andy, A., Ramesh, M., & Tharayil, A. S. (2025). Virtual labs and simulation tools: Enhancing STEM education. In *Revolutionizing education with remote experimentation and learning analytics* (pp. 607–632). IGI Global Scientific Publishing. https://doi.org/10.4018/979-8-3693-8593-7.ch035

Kalyani, D. (2024). The role of technology in education: Enhancing learning outcomes and 21st century skills. *International Journal of Scientific Research in Modern Science and Technology, 3*, 05–10. https://doi.org/10.59828/ijsrmst.v3i4.199

Khanduri, V., & Teotia, D. (2023). Revolutionizing learning: An exploratory study on the impact of technology-enhanced learning using digital learning platforms and AI tools on the study habits of university students through focus group discussions. *International Journal of Research Publication and Reviews, 4*, 663–672. https://doi.org/10.55248/gengpi.4.623.44407

Koehler, M. J., Mishra, P., Kereluik, K., Shin, T. S., & Graham, C. R. (2014). The technological pedagogical content knowledge framework. In J. M. Spector, M. D. Merrill, J. Elen, & M. J. Bishop (Eds.), *Handbook of Research on educational communications and technology* (pp. 101–111). Springer. https://doi.org/10.1007/978-1-4614-3185-5_9

Koh, J. H. L., & Kan, R. Y. P. (2021). Students' use of learning management systems and desired e-learning experiences: Are they ready for next generation digital learning environments? *Higher Education Research & Development, 40*(5), 995–1010. https://doi.org/10.1080/07294360.2020.1799949

Kumar, D., Haque, A., Mishra, K., Islam, F., Kumar Mishra, B., & Ahmad, S. (2023). Exploring the transformative role of artificial intelligence and metaverse in education: A comprehensive review. *Metaverse Basic and Applied Research, 2*, 21. https://dialnet.unirioja.es/servlet/articulo?codigo=9869857

Kundu, A., & Bej, T. (2021). COVID-19 response: Students' readiness for shifting classes online. *Corporate Governance: The International Journal of Business in Society, 21*(6), 1250–1270. https://doi.org/10.1108/CG-09-2020-0377

Major, L., Francis, G. A., & Tsapali, M. (2021). The effectiveness of technology-supported personalised learning in low- and middle-income countries: A meta-analysis. *British Journal of Educational Technology, 52*(5), 1935–1964. https://doi.org/10.1111/bjet.13116

Matthew, U., Kazaure, J., & Okafor, N. (2018). Contemporary development in E-Learning education, cloud computing technology & internet of things. *EAI Endorsed Transactions on Cloud Systems*, 169173. https://doi.org/10.4108/eai.31-3-2021.169173

McHaney, R. (2023). *The new digital shoreline: How Web 2.0 and millennials are revolutionizing higher education*. Taylor & Francis.

Mohajan, H. K. (2020). Quantitative research: A successful investigation in natural and social sciences. *Journal of Economic Development, Environment and People, 9*(4), 50–79. https://www.ceeol.com/search/article-detail?id=939590

Mokonyane, M. M., Moodley, K., & Van, R. L. (2021). Jasperactive as a gamification tool for motivation and engagement in adult learners. *Progressio, 42*(1), 1–18. https://doi.org/10.25159/2663-5895/10352

Nouwens, M., & Klokmose, C. N. (2021, June). A survey of digital working conditions of Danish knowledge workers. In *The 19th European Conference on Computer-Supported Cooperative Work. European Society for Socially Embedded Technologies (EUSSET)*.

Ogange, B. O., Agak, J. O., Okelo, K. O., & Kiprotich, P. (2020). Student perceptions of the effectiveness of formative assessment in an online learning environment. *Open Praxis, 10*(1), 29–39. https://doi.org/10.3316/informit.423669258504414

Ohei, K., Mantzaris, E., Ntshangase, B. A., & Olutade, E. O. (2023, September 1). *Incorporating new technologies into teaching in South Africa. | EBSCOhost*. https://doi.org/10.20525/ijrbs.v12i6.2715

Oke, A., & Fernandes, F. A. P. (2020). Innovations in teaching and learning: Exploring the perceptions of the education sector on the 4th industrial revolution (4IR). *Journal of Open Innovation: Technology, Market, and Complexity, 6*(2), 31. https://doi.org/10.3390/joitmc6020031

Parekh, R. (2025). *Principles of multimedia.* CRC Press.

Rahimi, R. A., & Oh, G. S. (2024). Rethinking the role of educators in the 21st century: Navigating globalization, technology, and pandemics. *Journal of Marketing Analytics, 12*(2), 182–197. https://doi.org/10.1057/s41270-024-00303-4

Richardson, B. (2023). *Decoding Public Perception: A Sentiment Analysis of the Top 10 Technology Trends That Will Shape the Next Decade. 16*, 1–26.

Rutkauskaite, R., Koreivaite, M., Karanauskiene, D., & Mieziene, B. (2022). Students' skills and experiences using information and communication technologies in remote physical education lessons. *Sustainability, 14*(23), Article 23. https://doi.org/10.3390/su142315949

Sajja, R., Sermet, Y., & Demir, I. (2025). End-to-End deployment of the educational AI hub for personalized learning and engagement: A case study on environmental science education. *IEEE Access, 13*, 55169–55186. https://doi.org/10.1109/ACCESS.2025.3554222

Sandberg, J., Holmström, J., & Lyytinen, K. (2020). Digitization and phase transitions in platform organizing logics: Evidence from the process automation industry. *Management Information Systems Quarterly, 44*(1), 129–153. https://urn.kb.se/resolve?urn=urn:nbn:se:umu:diva-168603

Sever, S. D., Tok, E., & Sellami, A. L. (2025). Sustainable development goals in a transforming world: Understanding the dynamics of localization. *Sustainability, 17*(6), Article 6. https://doi.org/10.3390/su17062763

Shambare, B., & Simuja, C. (2024). Unveiling the TPACK pathways: Technology integration and pedagogical evolution in rural South African schools. *Computers and Education Open, 7*, Article 100206. https://doi.org/10.1016/j.caeo.2024.100206

Shatrı, Z. G. (2020). Advantages and disadvantages of using ınformation technology in learning process of students. *Journal of Turkish Science Education, 17*(3), Article 3. https://www.tused.org/index.php/tused/article/view/1099

Shing, C. L., Saat, R. M., & Loke, S. H. (2018). The knowledge of teaching â€" pedagogical content knowledge (PCK). *MOJES: Malaysian Online Journal of Educational Sciences, 3*(3), Article 3. https://mojes.um.edu.my/index.php/MOJES/article/view/12781

Shulman, L. S. (1986). Paradigms and research programs for the study of teaching. In M. C. Wittrock (Ed.), *Handbook of research on teaching* (3rd ed.). Macmillan.

Sinha, S., Mazwane, S., Makhura, M. N., & Toit, A. du. (2024). *Chapter 22: sustainable development goal 9: Industry, innovation and infrastructure.* https://www.elgaronline.com/edcollchap/book/9781839105449/book-part-9781839105449-27.xml

Smallman, C., & Ryan, P. (2023). Transparency in governing technology enhanced learning. In *Technology-Enhanced learning and the virtual university* (pp. 43–57). Springer. https://doi.org/10.1007/978-981-99-4170-4_3

Stannard, R., & Sallı, A. (2019). *Using screen capture technology in teacher education.* Routledge.

Suleimanova, O., Guseinova, I., & Vodyanitskaya, A. (2020). The learning and educational potential of digital tools in humanities and social science. Society. integration. education. *Proceedings of the International Scientific Conference, 4*, 657–669. https://doi.org/10.17770/sie2020vol4.4851

Thangavel, S. (2025). *Revolutionizing education through augmented reality (AR) and virtual reality (VR): Innovations, challenges and future prospects* (SSRN Scholarly Paper No. 5195697). Social Science Research Network. https://doi.org/10.2139/ssrn.5195697

Tuong, N. K., & Dan, T. C. (2024). A study on duolingo mobile applications to improve efl students' listening comprehension performances. *European Journal of Alternative Education Studies, 9*(1), Article 1. https://doi.org/10.46827/ejae.v9i1.5342

Upadhyaya, D. H. (2024). *Digital education and economic transformation: Bridging the gap.* Meadow Publication.

van der Vyver, A. G., & Marais, M. (2015). Evaluating users' perceptions of the digital doorway: A narrative analysis. *Information Technology for Development, 21*(1), 99–112. https://doi.org/10.1080/02681102.2013.841629

Wang, T., Lund, B. D., Marengo, A., Pagano, A., Mannuru, N. R., Teel, Z. A., & Pange, J. (2023). Exploring the potential impact of artificial intelligence (AI) on international students in higher education: Generative AI, chatbots, analytics, and international student success. *Applied Sciences*, *13*(11), Article 11. https://doi.org/10.3390/app13116716

White, G., Wimmer, H., & Powell, L. M. (2020). *Pandemic shift: Impact of Covid-19 on IS/Microsoft Office specialist excel certification exam classes: Remote testing and lessons learned.*

Yalova, K., Babenko, M., Babenko, Y., Polyakova, N., & Anna, Z. (2025). Virtual laboratory as a tool to empower IT students. In E. Smyrnova-Trybulska, N.-S. Chen, P. Kommers, & N. Morze (Eds.), *E-Learning and enhancing soft skills: Contemporary models of education in the era of artificial intelligence* (pp. 259–277). Springer Nature Switzerland. https://doi.org/10.1007/978-3-031-82243-8_15

Chapter 9
Design and Implementation of a Smart Client–Server Attendance Register for Academic Classrooms to Minimize Latency and Enhance Flexibility

Hendrick Musawenkosi Langa and Rabelani Dagada

Abstract Taking an attendance register is an important procedure in meetings. In a class session, it is mandatory for some lecturers to take an attendance register, as this is one of the important documents for quality assurance. Other important information can be derived from the attendance register, such as the performance of a student in a module relative to his or her rate of attendance. Therefore, monitoring and analyzing attendance data can provide crucial clues about the behavior of attendees and information about them which can be useful and helpful to decision-makers. The purpose of this study was to develop and design a system that minimizes the amount of time spent generating an attendance register. It answered two questions: How can latency be reduced, and how can the system be designed to be reconfigurable? Software engineering approaches and methods were adopted for the development of the client–server system. In this study, a concise and user-friendly solution is presented. In this solution, the attendees, namely students, were able to register themselves using their own devices. After registering, they were able to connect to the server and sign the attendance registers. The system latency was drastically reduced. In one measurement, the first client signed the attendance register at 09:37:54 a.m. and the last client signed at 09:45:58. The total time spent taking the attendance register was 8 min and 4 s. The other objective achieved was to develop and implement the capability of a reconfigurable device, that is, a device that acts as both a client and a server. This solution eliminated the use of pen and paper to take attendance. Future studies on the use of computer vision for implementing the system can be explored.

Keywords Client- server · Threading · Database · Attendance register · System latency · UML class diagrams · Multi-threading · Client · Server · Use case · Digital pen · Inter-process communication · Database query

---

H. M. Langa (✉)
School of Computing, Center for Augmented Intelligence and Data Sciene, University of South Africa, Pretoria, South Africa
e-mail: langahm@unisa.ac.za

R. Dagada
Graduate School of Businesss Leadership, University of South Africa, Pretoria, South Africa

K. Dube and L. M. Tjabane (eds.), *Advancing Sustainable Development Goals in Higher Education*, https://doi.org/10.1007/978-3-032-27646-9_9

9.1 Introduction and Background

The old and traditional way of taking an attendance register is by using a pen and paper (Iio, 2016). Unfortunately, this method is still used today in many instances because there are user-friendly and effective solutions available for performing this important task. Calling the names of attendees one by one and ticking them off if they are present is clearly a waste of time (Makua et al., 2023). It is estimated that it takes approximately 10 minutes to take an attendance register of a small class. It is easy to imagine how much more time is lost when the class groups are large. In fact, it is impossible to take attendance registers for more than 500 attendees. This leaves almost no time for the purpose of the meeting. Unfortunately, this method is very inefficient in many ways. First, it is time-consuming, labor-intensive, and resource-intensive. Second, for large classes, for instance, 1500 students, it would be impossible to take an attendance register in one hour. It would also consume a lot of paper, which is wasteful. This method has since been challenged by modern methods since the inception of computers and technology. This process requires monitoring by an administrator to ensure that the register is completed accurately. The attendance register system used in a university is assumed to have two types of users: teachers or lecturers and students or learners (Zhang et al., 2019).

Furthermore, taking an attendance register is not limited to an academic environment. Many other settings warrant this practice, and consequently, there is a need for an efficient and effective method to do so. This article presents an efficient and effective solution for all spaces where an attendance register is required. The effectiveness of this solution is measured by the reduction in the latency of the system. Latency refers to the time required to complete an attendance register. Second, this study presents a reconfigurable system. In this design, a client can be a server and vice versa. Any of the devices can be a database server, and all devices are regroupable. This implies that a device can belong to an unlimited number of groups. For example, a student enrolls for more than one module; therefore, they belong to a certain number of module groups. Furthermore, a student can belong to several practical groups simultaneously. Therefore, database group attributes can be used to identify groups.

Attendance records serve as an important discipline tool used by many organizations (Johar et al., 2018). Hence, many researchers have made invaluable contributions to this field. However, further improvements are necessary because of the inefficiencies of these modern techniques. The lecturer, for example, can enforce attendance of his student by allocating a certain percentage toward the final mark of the student. This encourages students to attend lectures. The use of mobile devices by university and college students has increased over the years (Navin et al., 2017). This trend is unavoidable and should be embraced as we live in the era of the fifth industrial revolution (5IR). Less industrialized nations are still lagging behind in embracing and using the latest technologies, such as electronically taking attendance registers.

The approach in the development of these mobile systems means that designers need to analyze the system requirements and create mobile learning applications that can be used inside and outside the classroom to enhance effective teaching and learning as well as usage in other organizations. Therefore, it is easy to see that taking an attendance register has its own modalities and complications. A rudimentary practice of taking an attendance register, that is, using a piece of paper and a pen, is replaced by modern systems, the design of which requires that the following issues be addressed:

Latency: When the attendance register is generated, it is necessary to spend minimal time generating it. Several requests are sent to the server, which responds. The latency of the system determines the speed at which each request is processed (Shah et al., 2016). The sum of all requests typically becomes the total latency for completing the task of generating an attendance register. Multithreading is a necessary solution to prevent clashes when clients simultaneously send requests. In multithreading, the server can handle multiple requests simultaneously without crashing. With the traditional pen-and-paper method, it would take a few minutes for an attendee to complete the registration form. If information about that client was already captured, it would take one or two minutes to locate his name in the list and then sign off. The total time taken to complete the signing of the register has its own latency. The total latency of completing the whole register for n clients will be n times the average time it takes one client. Threads are useful for reducing latency in the system; therefore, their incorporation is required.

Storage: One of the most common forms of data storage is an online database (Murali et al., 2024). This allows the application to be developed as a web application. A local database storage system would also work well. Because the database only stores information as text, there is a huge reduction in data storage requirements. No files, images, or other high-format data are required. In this case, the system is relatively efficient and effective in terms of data storage. Although the system effectively accomplishes its goal of effectively storing data required for a register, this does not prohibit the storage of data in other formats, as may be required by other applications. Large numbers can be stored which may not require huge storage centers.

Client–Server/Server–Client Architecture: The elegance of the system presented in this study is that any device can be both a client and a server. The device can be a desktop, laptop, or mobile device. If the device is configured as a client, it hosts its own database. Other machines configured as clients can connect and sign a register which the server machine can generate. All they require are login credentials to enable the signing of an attendance register.

Client Groups: Another advantage of the system is that clients can belong to an unlimited number of groups in the database. The administrator can create and classify clients using a group ID. For example, a client can belong to a group of researchers, PAs, engineers, accountants, tutorial groups, class groups, and so on.

Research Questions: All research questions were answered. The following questions were addressed:

- How can a network-based attendance registration system be implemented?
- The time required to complete a register is usually unnecessarily long. How can the time spent on taking a register be reduced? In other words, how can the latency be reduced?
- How can the system be reconfigured to be a server as well as a client?

Hypothesis: It is therefore hypothesized that if each client can have their own client device and software, they can sign a register by themselves rather than being attended to by an administrator who is expected to help each client sign in a register one by one. This is referred to as client–server computing. Client–server computing systems addressed the issue of latency in the process and reduced the time spent in signing a register. In fact, the time it takes to complete an attendance register can be determined by the administrator or lecturer. How is this done? The administrator simply turns the server on and off at their discretion. The queue and cost problems were identified by Iio (2016) are major problems in the design of such systems. Therefore, this study attempts to solve the queue problem by minimizing latency as much as possible. The server administrator provides access to the client to log in and sign the attendance register. When the client has signed, the server administrator can close or disconnect the server.

Therefore, the hypothesis as illustrated graphically in Fig. 9.1 is that if there are more devices, that is, clients, the time taken to complete the attendance register will be reduced. The following graph illustrates the hypothesis.

In practice, the attendance register can be taken as $t = 0$, that is, the start of the meeting to $t =$ maximum time which is the length of the meeting. It may not make sense to take the attendance register after that time, since the meeting may be over. The graph above shows an inverse relationship between the number of clients and the time taken to complete the attendance register. The optimal number of clients is equal to the number of participants. Any number of clients greater than the attendance also makes sense since only the attendees are allowed to sign the attendance register.

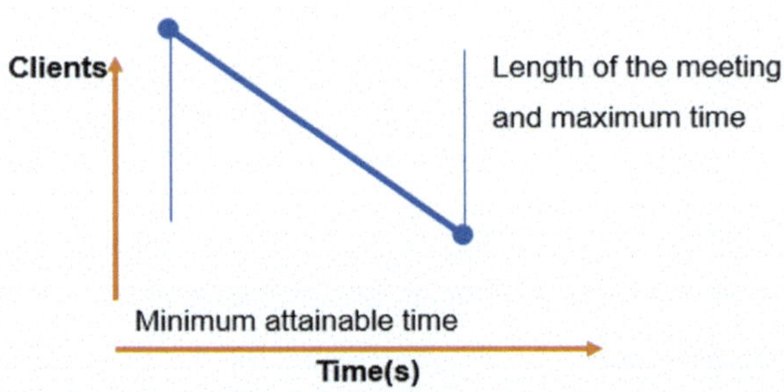

Fig. 9.1 Graphical illustration of the hypothesis statement

9.2 Literature Review and Related Studies

In academic institutions, maintaining an attendance register is mandatory. The data generated from this exercise are useful for decision-making and the promotion of teaching and learning quality. The development of the client server system is motivated by the fact that the attendance of students has some impact on their performance (Othman et al., 2009). There have been many attempts and approaches to address the matter of taking an attendance register which is key to improving the quality of education (Gatsheni et al., 2007). In many schools and colleges, teachers keep records of student attendance and later enter the percentage of student attendance into a computer (Sunehra & Goud, 2016). Educators have grappled with the cumbersome manual processes of paper-based attendance sheets for many years, whereas electronic-based systems have somewhat helped to improve the taking of attendance registers. This results in more manpower requirements and hence is more costly; therefore, a proper attendance management system is crucial for an academic institution to disseminate and ensure quality education to every student (Akbar et al., 2018). However, a few institutions still use an attendance register system in their meetings or classes. Technophobia also plays a significant role in preventing potential users from using and embracing new technology. Technophobia is a phenomenon that is intertwined with new technologies, and as long as there are new technologies, there will be phobias associated with them (Khasawneh, 2018).

A QR Code (abbreviated Quick Response Code) is a trademark for a type of matrix barcode designed for the automotive industry in Japan (https://citeseerx.ist.psu.edu/document?repid=rep1&type=pdf&doi=288f0459675d41e2d3bbb8b6b65bc927ffe57262). The QR Code-based technique is a mechanism for registering attendance remotely and is meant to prevent unauthorized usage of the system (Navin et al., 2017). One of the main challenges with taking an attendance register is that it is time-consuming; even web-based systems take a long time to complete this routine task (Agarwal et al., 2019). The QR Code can be displayed so that when the students enter the classroom, they can take out their smartphones and scan the QR Code (https://citeseerx.ist.psu.edu/document?repid=rep1&type=pdf&doi=288f0459675d41e2d3bbb8b6b65bc927ffe57262). However, this system has a serious deficiency in terms of latency if the class is large.

Although web-based systems are a preferred technology (Othman et al., 2012), there are advantages and disadvantages to using such technologies. In this study, a network-based approach was used to implement the system, which also has its own advantages and disadvantages. Web-based systems are known for their efficient time saving (Othman et al., 2012). Dankar and Kundapur (2019) argue that other systems have loopholes that allow students to cheat the system by signing for other absent students, and these systems are also resource-intensive.

Face recognition is a recent technology that is widely used. Face recognition is characterized by two steps. The first step was face detection. Second, the detected face is compared with the existing faces in the database by comparing various features, such as the eyes and nose (Nadhan et al., 2022). Face recognition has been one of the most active areas of computer vision and applied pattern recognition (Shoja Ghiass et al., 2014). Although face recognition systems have become more prevalent in recent years, they are not as common in South African universities. Some biometric methods include voice and iris recognition (Zhang et al., 2019). Face recognition has been used for both the verification and identification of students in a classroom (Ahire et al., 2021). One major drawback of this technology is the large computational task involved in detection and identification. The system relies on cloud resources to improve the facial identification performance in large video databases. A cloud is a pool of easily usable and accessible virtualized resources, such as development tools, hardware, and services. These resources can be dynamically reconfigured to meet several users' needs simultaneously and to adjust to variable loads (Ahire et al., 2021).

Fingerprint attendance technology automates the attendance procedure in educational institutions using biometric data. Fingerprints are a form of biometric identification which do not change throughout one's life (Akinduyite et al., 2013). In this system, the client or student must be enrolled, after which they can authenticate using fingerprints as a signature. It can work well in cases where the number of clients is not large, but in large classes, it can be slow in completing the attendance register, as mentioned earlier.

Although a web-based system is a preferred technology (Othman et al., 2012), there are advantages and disadvantages to using such technologies. In this study, a network-based approach was used to implement the system which also has its own advantages and disadvantages. Web-based systems are known for their efficient time saving (Othman et al., 2012). Dankar and Kundapur (2019) argue that other systems have loopholes that allow students to cheat the system by signing for other absent students, and these systems are also resource-intensive.

A web-based student attendance system using Radio Frequency Identification (RFID) is not a new concept in the design and implementation of an attendance register. The student card is embedded with an FID tag which is read by the reader (Mohammed, 2013).The FRID system is then connected or linked to a database using a device such as a computer or an electronic circuit. The FRID reader sends a pulse of radio energy to the tag and then listens to the tag's response. The information is processed and sent to the web server which sends it to the database for storage. The attendance register can be stored and updated. The shortcoming is that students have to queue to log in their attendance which is again time-consuming.

Firebase is a tool on the Google platform for building real-time database systems. The Firebase database helps to sink and store data in real time. Real-time database systems manage their data (e.g., temperature and aircraft location) in application-dependent structures (https://citeseerx.ist.psu.edu/document?repid=rep1&type=pdf&doi=288f0459675d41e2d3bbb8b6b65bc927ffe57262). Firebase is a modern technology that is much faster than Relational Database Management System (RDMS) for managing random data (Khawas & Shah, 2018). Data in Firebase are stored as JSON and remain available even if the system is offline. The Firebase database infrastructure is managed by Firebase, and it takes a few lines of code to implement the system. Firebase is a secure platform.

Attendance records serve as an important discipline tool that many organizations use (Johar et al., 2018); hence, many researchers have made invaluable contributions to this research area. However, more improvements are necessary due to inefficiencies of these modern techniques. The use of mobile devices by university and college students has increased over the years (Navin et al., 2017). This trend is unavoidable and should be embraced as we live in the era of the fourth industrial revolution.

The approach in the development of these mobile systems means that designers need to analyze the system requirements and create mobile learning applications that can be used inside and outside the classroom to enhance effective teaching and learning as well as usage in other organizations. Therefore, it is easy to see that taking an attendance register has its own modalities and complications. A rudimentary practice of taking an attendance register, that is, using a piece of paper and a pen, therefore, requires improvements which are achieved by addressing the following issues:

9.3 Methodology

The design of the software system hinges on the following software engineering protocols and principles. These design principles include the use of UML diagrams, such as activity, use, and sequence diagrams. Software engineering involves communicating complex ideas simply using UML diagrams.

In this study, attendance register systems were designed as client-server and server-client systems. Any device can assume the client or server role. The server device is responsible for data storage, data processing, communication protocols, preparation of relevant reports, execution of client queries, management of connection protocols, and management of current registers. Figure 9.2 shows the use case diagram of the client-server attendance register system.

The development of the client server system is motivated by the fact that the attendance of students has some impact on their performance (Othman et al., 2009). There have been many attempts and approaches to address the matter of taking an attendance register which is key to improving the quality of education.

A typical client–server system is shown in the following Fig. 9.3. The client–server architecture allows several clients to send requests to the server. The server can then respond to requests made by the client. The server can also manipulate the database by performing CRUD operations on the database, that is, create, read, update, and delete operations.

Client processes should be able to launch all at once, whereas the server should be able to handle all clients simultaneously. This is accomplished using a technique known as multithreaded processing. This consequently means that all attendees should be able to log in and sign the attendance register simultaneously, whereas the server addresses the requests that are received from the client devices. Although only three client devices are shown in the figure above, as many client devices as required can be connected to the server device or stored on other devices.

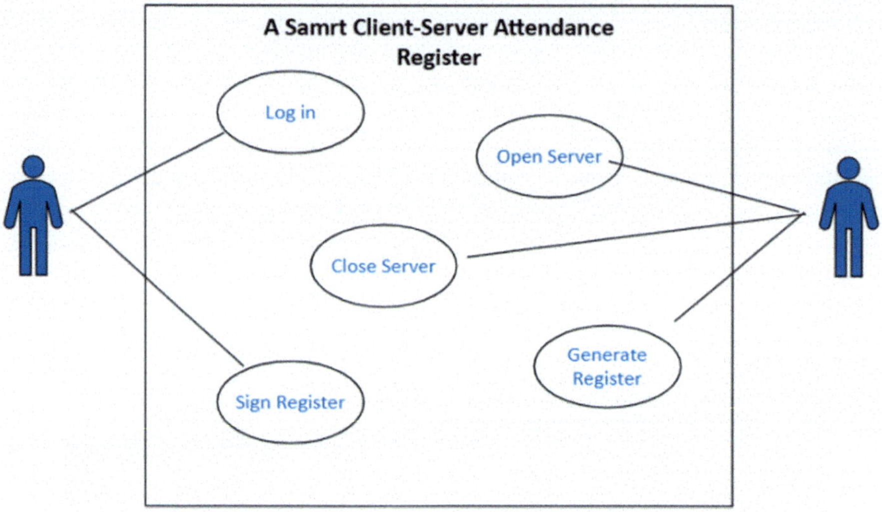

Fig. 9.2 UML diagram of the use case for the client–server attendance register system

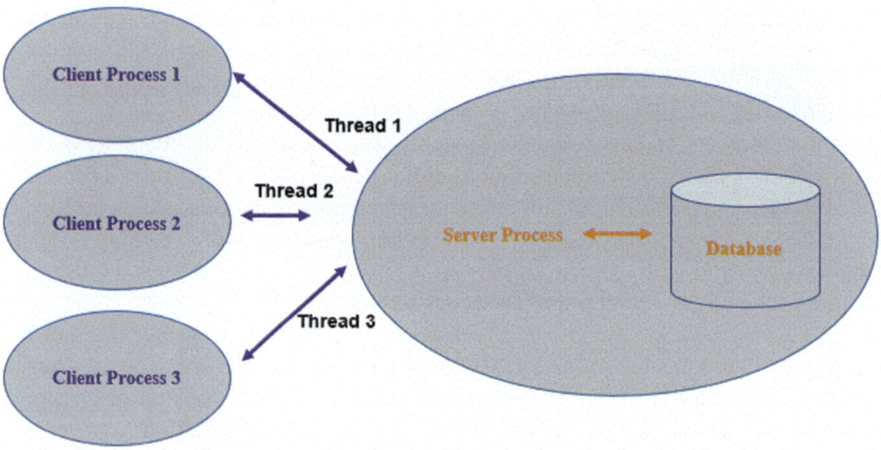

Fig. 9.3 Client–server system with a local database

The client device is a digital pen that can be used to sign a register remotely. Before a client can sign in for a meeting, they must be given access rights to the meeting. Once this access right is given, the user is required to authenticate, after which the client can sign in. Assuming a large number of clients attending a particular meeting, they can all register simultaneously. Therefore, there is no time lag caused by queuing in the traditional registration process because threading technologies are applied during system design.

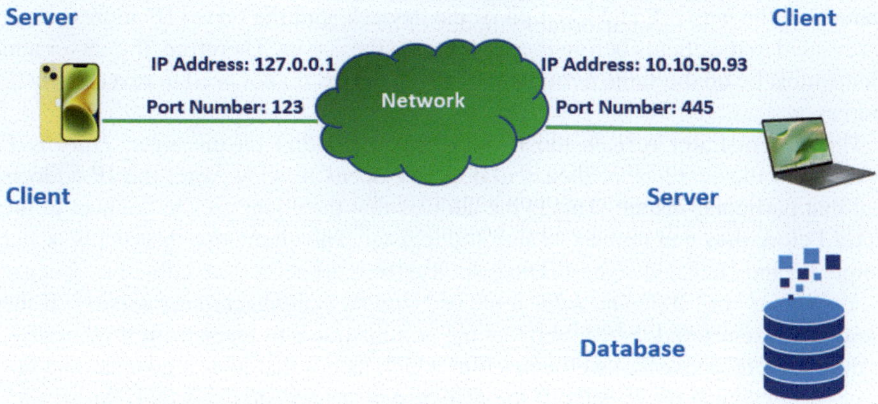

Fig. 9.4 Basic model: attendance register architecture

The client and server must be on the same network to establish communication. If the client is on a different network from the server, communication is not possible. The second requirement is that the client should authenticate to establish a connection with the server which is listening at a particular IP address at a particular port number for client requests. Any number of clients can connect simultaneously and make requests to sign the attendance register.

Figure 9.4 shows the basic architecture of the client-server system. What is unique about it is that any client machine can be configured as a server, and vice versa. Any number of clients can send requests to the server at any time as long as they are on the same network and successfully authenticate. The set of clients can be defined as follows:

$$\text{clients} = \{C_1, C_2, C_3 \ldots, C_n\}, \tag{9.1}$$

where

$C_i \in$ clients is the client device.

The cardinality of clients is variable and is given by

$$|\text{clients}| = N, \tag{9.2}$$

where

N is set of natural numbers.

Therefore, the client should have both client and server software installed. Each device will have its own database for its own purposes.

A. Dynamic IP Addresses

Figure 9.4 shows fictitious IP addresses; however, real IP addresses are allocated dynamically. Because the system is network-based, the IP addresses change with changes in the network and location of the devices, and the server IP address must be revealed to the clients before they can access the server. Therefore, the server and clients must be on the same network. However, clients do not need to reveal their IP addresses.

The administrator reveals the IP address by clicking on the alias: *Trackie/IP* Address on the server app, then a dialog box opens to show what the IP address is at that particular instant. This is the address that the client uses to connect to the server before they can request to sign the register. The client also inserts his or her unique ID and clicks on sign in. The response from the server can either be "*You are not registered*" or "*Welcome John Joe!*" or "*Unable to connect to the server*." If the client is not registered, he can request the administrator to register him; otherwise, he may get a response that welcomes him which means that he is registered and has successfully signed the register. If the response is "*Unable to connect to the server*" then the client is not logged into the network.

B. First Time Entries to the Database

As in any database system, the entry of first-time clients occurs once when the client is being registered. The shortcoming of first-time registration of clients is that it can be time-consuming; therefore, a functionality to read files that may contain information about clients may be required unless there is no such list of clients.

C. Definition of Variables

In the design and development of the client-server system, the initial development relates to the design of the server and its functionalities. The server should be designed such that it is capable of generating an attendance register locally. This process is slow because each attendee should be first registered before he or she can sign the attendance register. Only one device is used as an attendance register to sign in one attendee at a time.

D. Data Collection and Analysis

Once the design of the client server was completed, quantitative data collection was performed using the system. The administrator prepares the server to collect clients who sign the register. The number of clients who sign in is known in principle because the client has already been registered in the system. The exact number of attendees will be known after the registration process is completed.

9.3.1 Data Collection

Data collection can be implemented using a software program (Bourenkov & Popov, 2006). In this study, the client-server software itself was used to collect data which were analyzed later. At this stage, clients can sign the attendance register because they have the client application on their devices, as an attendance register to sign in one attendee at a time. With the application on their devices, they can make requests

Table 9.1 Database table of the clients

Attribute	Value
Client number	********
Client name	********
Client group	********
Cell number	********
Email	********
Date	********
Time	********

to the server as soon as a connection is established. The server listens to the clients to accept or reject them. When a client is accepted, it can make a request to which the server responds. The data contain the following attributes as shown in Table 9.1.

Other attributes can be added or removed, but the most important of these are the number (it could be a cell number, a staff number, a student number, an identity number, or any number that is unique), name, group, and date time stamp.

9.3.2 Data Analysis

The date time stamp contains crucial information about the date that the client signed an attendance register and the time at which it was signed. The optimal outcome is obtained when the clients sign the register on the same date and time. This is not exactly practical in practice, but the closer the date time stamps are to one another, the closer the outcome is to the minimum time spent in creating an attendance register. Restricting the time taken to complete an attendance register is an essential tactic to minimize the time taken to generate an attendance register. This implies that the server should be turned on and off at known times.

9.3.3 Multithreading

Multithreading also helps minimize time loss when requests are made by several clients. Moreover, different students or clients can authenticate at different times, and they can send requests to sign a register at different times. All these activities are meant to occur within a defined time slot of less than the meeting time as shown in Fig. 9.5. Multithreading serves the purpose of managing queuing clients.

The server software and database reside on the server device which can be a computer, smartphone, or cloud database. The register can be taken using only the server device, but it is sequential. Sequential attendance register taking is very slow, but it eliminates paper sheets. This method is sufficient for small groups and classes.

Fig. 9.5 Latency triangle

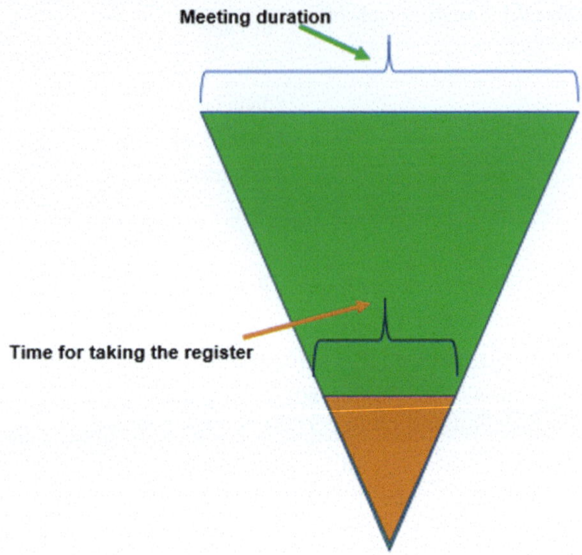

The client software resides on client machines such as smartphones and computers. This completes the client–server system design. All client devices can also be configured as servers with their own databases, and all server machines can be configured as client machines. This is a novel approach that makes the attendance registration system adaptable and versatile.

The time required to complete the process of taking an attendance register ranges between 0 and infinity. Under normal circumstances, the time taken to complete an attendance register is in the order of minutes or even seconds. Effective class register management requires fast and accurate approaches (Mohammed et al., 2018). This approach is fast, accurate, and reliable for use in any environment beyond educational institutions.

The acceptable latency level is on the order of milliseconds (Ritter et al., 2015). The response times of technical systems have long been a subject of debate. Shorter response times provide more efficiency from the user's perspective, whereas longer response times may be considered unattractive and may even prohibit the user from completing a task efficiently and successfully. In this study, the process latency of the attendance register is related to how long it takes to complete the register, given the duration of the meeting. The time it takes to complete a register in comparison to the time it takes to conduct a meeting is given by the following inequality:

$$t_r \leq t_d, \tag{9.3}$$

where

t_r is the time it takes to complete an attendance register.

t_d is the duration of the meeting or length of the class period.

The value of t_r is ideally supposed to much smaller than the value of t_r in general than the duration of the meeting or length of the class period. While the duration of the meeting may be predetermined, and the user may or may not have control over it, the user should have considerable or total control over the time it takes to complete the attendance register.

9.4 Design and Implementation

The design and implementation of the system were accomplished using Python, Android, and Java platforms. These are modern tools available; however, platforms such as Django, Java script, CSS, and HTML may be required when developing web applications.

9.4.1 Client Side App

The client app resides on the client device and is used to sign the attendance register. This is the "digital pen" that the attendee connects to the server. It is essential to be on the same network as the server machine before an attempt to connect can be made, as indicated previously. Figure 9.6 shows a client app, "*Trackie Client*" that each attendee should have on their devices.

Once the client is authenticated on the same network as the server, the client can attempt to make a request to sign the register. If the request was successfully processed, the server device notified the client device that the execution of signing a register was completed successfully. The server then updates the attendance register database.

9.4.2 Server Side App

The server is responsible for the logic and hosting the database. CRUD operations are all performed by the server API, that is, creation, reading, updating, and deletion of information are processed on the server app as shown in Fig. 9.7.

Other functionalities, such as searching for the IP address of the current network and starting and stopping the server, are also performed from the server side. Reports were also processed on the server side.

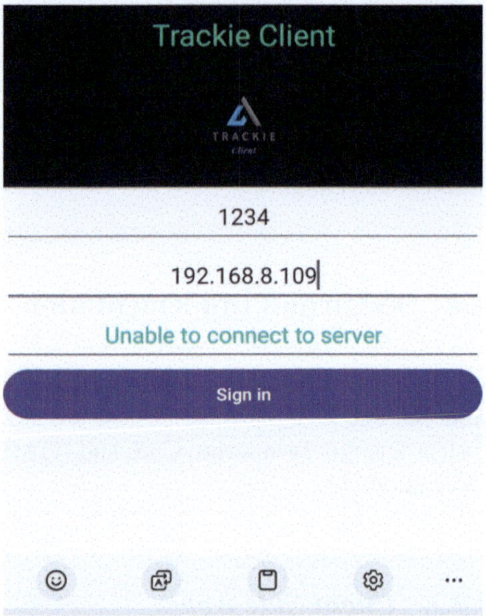

Fig. 9.6 Trackie client—a digital pen

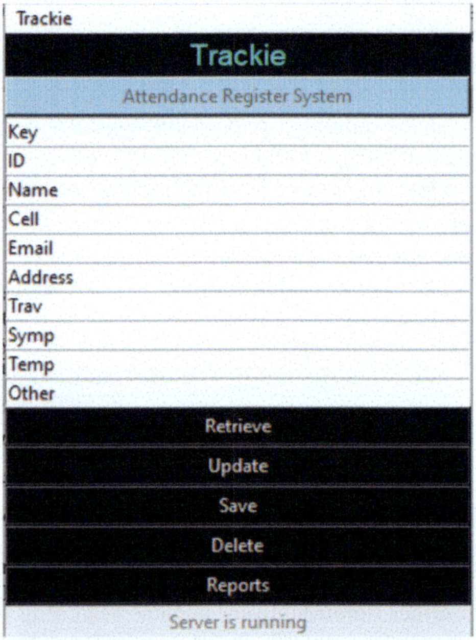

Fig. 9.7 Trackie—the server

9.4.3 *Queries and Database Searches*

Queries and database searches for information are crucial for the effectiveness of the server. The ability to query the database is a useful capability where the administrator or lecturer can search for a client using any criterion, that is, by name, group, email address, title, address, ID, or any other criteria that he deems fit.

9.4.4 *The Attendance Register Reporting*

While the clients sign the attendance register, the server automatically generates the attendance register in the background. A unique feature of this system is that the attendance register lists are not permanently stored in the system which is a significant advantage because this technique eliminates the risk of having to store them in the device. The list is available for the day the attendees attended the meeting; the following day, that list will not necessarily be available to generate the same attendance register. Therefore, the list as shown in Fig. 9.8 is generated when clients sign, after which it can be stored elsewhere rather than in the current server, to prevent data explosion in the server.

The register is automatically generated on the interface. The administrator can retrieve and save the data for future use. The attendance register is saved as a PDF document at a location decided by the administrator on the server.

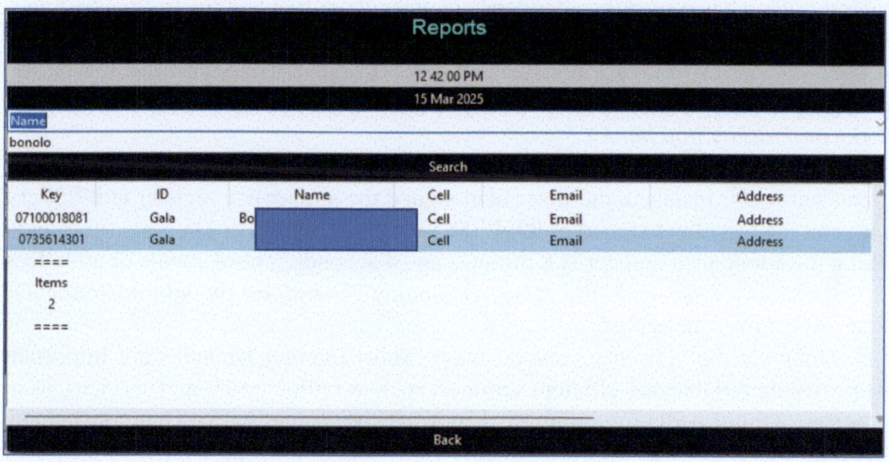

Fig. 9.8 Attendance register report screen

9.5 Results and Discussion

A new client is registered in the system using a unique number. This unique number can be a student number in the case of a university or college. In the case of employees of a company, it can be a staff number. It can be a cell number in general or any number as long as the number is unique to that client.

9.5.1 Attendance Register Report

After the client has signed the attendance register, the server administrator can immediately generate the attendance register. In fact, as the clients sign the attendance register, the register is automatically generated in the server app. Figure 9.9 shows an example of an attendance register used in the study.

In the case of students' attendance register, the lecturer can file the report for future use, such as using the register for the Engineering Council of South Africa (ECSA) file system purposes. In general, it can be a cell number or any other number as long as that number is unique to that client.

9.5.2 Latency in Signing the Attendance Register

After the client has signed the attendance register, the server administrator can immediately generate the attendance register. It is up to the administrator to decide when to open the server to take a register and when to close it. To start the server, the administrator simply clicks on *Trackie/Start Server*, and to stop the server, he simply clicks on *Trackie/Stop Server*.

Figure 9.10 shows the latency in signing the attendance register of the participants. In this particular instance, the first client signed the attendance register at **09:37:54 a.m.** and the last client signed at **09:45:58** as seen in Fig. 9.10. The total time spent taking the attendance register is 8 minutes and 4 seconds. These clients or attendees can sign at any time while the server is running. Therefore, the administrator has total control over the server.

Latency can be viewed in several ways. Short Internet latencies are important for providing reliable and efficient services, such as online goods and services, as in Amazon, or multimedia streaming, such as YouTube (Tolga, 2015). There is a latency that can be referred to as network latency. This is the time it takes between the client pressing the sign on their device and the response from the server confirming that the client has signed the register. The other form of latency can be viewed as the time it takes from the first client signing the register to the last client signing the register. Latency is the difference between the first client who signed the register and the last client. Latency can be defined as how quickly a client can be processed when they

Trackie Attendance Register

Report

30-Aug-2024

No	ID	Name	Contact	Date	Time
1	218128185	[redacted]		30 Aug 2024	09 45 48 AM
2	222337850			30 Aug 2024	09 43 35 AM
3	223176826			30 Aug 2024	09 43 13 AM
4	223214000			30 Aug 2024	09 44 24 AM
5	223361380			30 Aug 2024	09 45 36 AM
6	223522244			30 Aug 2024	09 37 26 AM
7	223771406			30 Aug 2024	09 42 41 AM
8	223954411			30 Aug 2024	09 37 54 AM
9	224144545			30 Aug 2024	09 38 38 AM
10	224145304			30 Aug 2024	09 38 02 AM
11	224198629			30 Aug 2024	09 45 21 AM
12	224295748			30 Aug 2024	09 38 22 AM
13	224367072			30 Aug 2024	09 45 33 AM
14	240251067			30 Aug 2024	09 42 15 AM
15	240254260			30 Aug 2024	09 45 16 AM
16	240359828			30 Aug 2024	09 41 01 AM
17	240391942			30 Aug 2024	09 38 20 AM
18	240711262			30 Aug 2024	09 39 02 AM
19	240755723			30 Aug 2024	09 44 58 AM
20	240934784			30 Aug 2024	09 43 42 AM

Fig. 9.9 Attendance register report

sign the register. The time required to process one client is instantaneous. Therefore, latency is negligible, whereas the time taken to process 20 students in this example is 8 min and 4 s. One can say that the latency in this case was 8 min and 4 s. This latency is dependent on several parameters, such as the predetermined time allocated to taking a register. This means that the administrator has the option of prescribing to the attendees or clients the amount of time allocated to signing a register. The amount of time allocated can be anything from 1 min to any length of time.

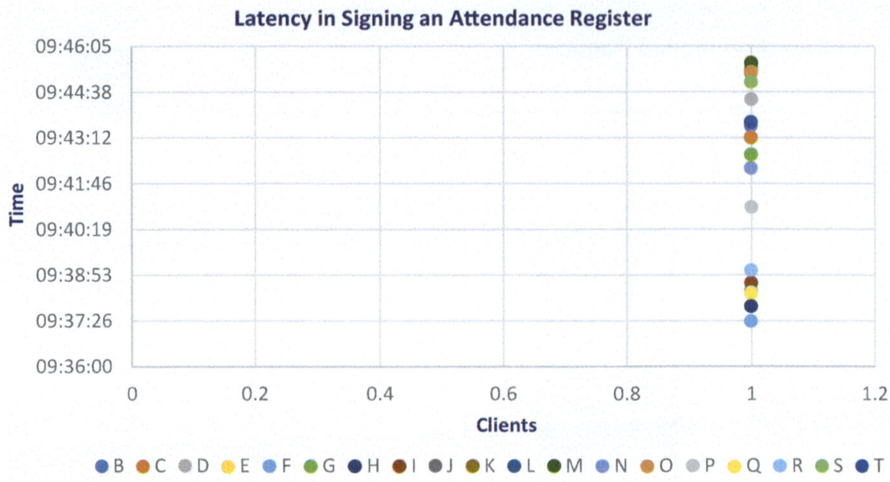

Fig. 9.10 Latency in Signing the Attendance Register

9.6 Conclusion and Recommendations

This study aimed to design and implement a smart client-server attendance register to address the use of paper and pen when taking an attendance register, reduce latency in generating an attendance register, implement a network-based system, and design a system that can allow users to configure their devices as both clients and servers.

The system was designed for a network environment rather than an Internet environment. It has its own advantages and disadvantages. It can also be redesigned for the Internet, which has its own advantages and disadvantages. Therefore, the implementation of the system, as well as a comparative study between a network design and an Internet or web design, is an opportunity for future research.

Acknowledgements The authors wish to thank the University of South Africa and Vaal University of Technology for moral and technical support and the DHET for their assistance with funding support while conducting research and for giving the authors the opportunity to publish their work in accredited research in conferences and journals.

References

Agarwal, P., Shukla, V. K., Gupta, R., & Jhamb, S. (2019). Attendance monitoring system through RFID, face detection and Ethernet network: A conceptual framework for sustainable campus. In *2019 4th International conference on information systems and computer networks (ISCON)*, pp. 321–325. https://doi.org/10.1109/ISCON47742.2019.9036209

Ahire, S., Jadhao, R., Gawai, V., & Giri, P. (2021). Automated student attendance management system using face recognition. *International Journal of Advanced Research in Science, Communication and Technology*, 187–192. https://doi.org/10.48175/IJARSCT-2019

Akbar, Md. S., Sarker, P., Mansoor, A. T., Al Ashray, A. M., & Uddin, J. (2018). Face recognition and RFID verified attendance system. In *2018 International conference on computing, electronics & communications engineering (iCCECE)*, pp. 168–172. https://doi.org/10.1109/iCCECOME.2018.8658705

Akinduyite, C. O., Adetunmbi, A. O., Olabode, O. O., & Ibidunmoye, E. O. (2013). Fingerprint-based attendance management system. *Journal of Computer Sciences and Applications, 1*(5), 100–105. https://doi.org/10.12691/jcsa-1-5-4

Bourenkov, G. P., & Popov, A. N. (2006). A quantitative approach to data-collection strategies. *Acta Crystallographica Section d: Biological Crystallography, 62*(1), 58–64. https://doi.org/10.1107/S0907444905033998

Dankar, A., & Kundapur, P. P. (2019). Automated mobile attendance system (AMAS). In *2019 International conference on advances in computing, communication and control (ICAC3)*, pp. 1–6. https://doi.org/10.1109/ICAC347590.2019.9036787

Gatsheni, B. N., Kuriakose, R. B., & Aghdasi, F. (2007). Automating a student class attendance register using radio frequency identification in South Africa. In *2007 IEEE international conference on mechatronics*, pp. 1–5. https://doi.org/10.1109/ICMECH.2007.4279977

https://citeseerx.ist.psu.edu/document?repid=rep1&type=pdf&doi=288f0459675d41e2d3bbb8b6b65bc927ffe57262

Iio, J. (2016). Attendance management system using a mobile device and a web application. In *2016 19th International conference on network-based information systems (NBiS)*, pp. 510–515. https://doi.org/10.1109/NBiS.2016.44

Johar, R., Qaisar, S. M., Subasi, A., & Kurdi, R. F. (2018). An event driven attendance tracker. In *2018 21st Saudi computer society national computer conference (NCC)*, pp. 1–6. https://doi.org/10.1109/NCG.2018.8592968

Khasawneh, O. Y. (2018). Technophobia: Examining its hidden factors and defining it. *Technology in Society, 54*, 93–100. https://doi.org/10.1016/j.techsoc.2018.03.008

Khawas, C., & Shah, P. (2018). Application of firebase in android app development-A study. *International Journal of Computer Applications, 179*(46), 49–53. https://doi.org/10.5120/ijca2018917200

Makua, M., Akinlolu, M., Sithole, M., Gumede, P., Nyondo, C., Khuzwayo, N., & Mhlongo, M. (2023). *Proceedings of the 10th focus conference (TFC 2023)*. Springer Nature.

Mohammed, A. A. (2013). Web-server based student attendance system using RFID technology. *International Journal of Engineering Trends and Technology*.

Mohammed, K., Tolba, A. S., & Elmogy, M. (2018). Multimodal student attendance management system (MSAMS). *Ain Shams Engineering Journal, 9*(4), 2917–2929. https://doi.org/10.1016/j.asej.2018.08.002

Murali, N., Rajesh, R., Sridharan, S., Mariadas, A. E. P., Murugan, S. P., & Manikandan, S. (2024). A GPS-based face attendance register system using android applications stored in the cloud. In *2024 11th International conference on computing for sustainable global development (INDIACom)*, pp. 17–22. https://doi.org/10.23919/INDIACom61295.2024.10498447

Nadhan, A. S., Tukkoji, C., Shyamala, B., Dayanand Lal, N., Sanjeev Kumar, A. N., Mohan Gowda, V., Adhoni, Z. A., & Endaweke, M. (2022). Smart attendance monitoring technology for Industry 4.0. *Journal of Nanomaterials, 2022*(1), 4899768. https://doi.org/10.1155/2022/4899768

Navin, K., Shanthini, A., & Krishnan, M. B. M. (2017). A mobile based smart attendance system framework for tracking field personals using a novel QR code based technique. In *2017 International conference on smart technologies for smart nation (SmartTechCon)*, pp. 1540–1543. https://doi.org/10.1109/SmartTechCon.2017.8358623

Othman, M., Ismail, S. N., & Noradzan, H. (2012). An adaptation of the web-based system architecture in the development of the online attendance system. In *2012 IEEE conference on open systems*, pp. 1–6. https://doi.org/10.1109/ICOS.2012.6417619

Othman, M., Ismail, S. N., & Raus, M. I. (2009). *The development of the web-based Attendance Register System (ARS) for higher academic institution: From feasibility study to the design phase.*

Ritter, W., Kempter, G., & Werner, T. (2015). User-acceptance of latency in touch interactions. In M. Antona & C. Stephanidis (Eds.), *Universal access in human-computer interaction. Access to interaction* (pp. 139–147). Springer International Publishing. https://doi.org/10.1007/978-3-319-20681-3_13

Shah, N. B., Lee, K., & Ramchandran, K. (2016). When do redundant requests reduce latency? *IEEE Transactions on Communications, 64*(2), 715–722. https://doi.org/10.1109/TCOMM.2015.2506161

Shoja Ghiass, R., Arandjelović, O., Bendada, A., & Maldague, X. (2014). Infrared face recognition: A comprehensive review of methodologies and databases. *Pattern Recognition, 47*(9), 2807–2824. https://doi.org/10.1016/j.patcog.2014.03.015

Sunehra, D., & Goud, V. S. (2016). Attendance recording and consolidation system using Arduino and Raspberry Pi. In *2016 International conference on signal processing, communication, power and embedded system (SCOPES)*, pp. 1240–1245. https://doi.org/10.1109/SCOPES.2016.7955639

Tolga, S. (2015). *Enabling real-time mobile cloud computing through emerging technologies.* IGI Global.

Zhang, H., Feng, X., Liu, H., Guo, P., Krishnamoorthy, S., & Zhang, C. (2019). Cloud-based class attendance record system. In *2019 IEEE 5th international conference on computer and communications (ICCC)*, pp. 283–287. https://doi.org/10.1109/ICCC47050.2019.9064482

Part V
Sustainability, Art, and Global Perspectives

Chapter 10
Role of Higher Education in Green Technology Revolution: A Global Perspective

Kaitano Dube, Sinikiwe Gwatidzo, and Lazarus Maleho

Abstract Universities play a critical role in championing the aspirations of the Sustainable Development Goals and ensuring that no one is left behind. The role of universities is even more imperative in addressing some of the critical challenges facing the world, such as clean energy security and climate change. Given this central role, this study examines the role of higher education institutions in the green technology revolution, which aims to ensure carbon neutrality. Using bibliometric analysis of 100 Web of Science publications published between 2016 and 2025, analysed through critical document analysis, this study found that universities play a central role in the green technology revolution. Amongst other things, universities are trusted to develop programmes which focus on green technology, conduct studies leading to technological advancement in sustainability, foster a culture of green campuses and are responsible for policy advocacy amongst other things.

Keywords Climate change · Mitigation · Clean energy · Carbon neutral · SDGs

10.1 Introduction

Over the past two decades, the world has been battling an increasing number and impact of global change events (Chen et al., 2018; Filonchyk et al., 2024). Such events include battling the impacts of increasing population, pollution (Chikodzi et al., 2021; Dong et al., 2024) and climate change (Dube, 2024a). Anthropogenic

K. Dube (✉)
College of Economic and Management Sciences, University of South Africa, Pretoria, South Africa
e-mail: dubekaitano@gmail.com

Emirates Aviation University, Dubai Academic City, Dubai, United Arab Emirates

S. Gwatidzo
Registrar Department, Great Zimbabwe University, Masvingo, Zimbabwe

L. Maleho
Faculty of Human Sciences, Vaal University of Technology, Vanderbijlpark, South Africa

K. Dube and L. M. Tjabane (eds.), *Advancing Sustainable Development Goals in Higher Education*, https://doi.org/10.1007/978-3-032-27646-9_10

climate change has been of particular concern to humanity given the fact that there has been persistent increase in greenhouse gas emissions over the past decades (Dube et al., 2025). This is despite the slight decline in greenhouse gas (GHG) emissions that was observed at the height of the COVID-19 pandemic due to the wide-scale disruptions caused by the disease which was punctuated by company shutdowns (Nguyen et al., 2025; Ray et al., 2022).

The increase in GHG emissions has been matched by an increase in the number of loss and damage events attributed to climate change (Badr & El-Shazly, 2024; Mechler et al., 2020). The combined impact of the pandemic and losses from climate change has reenergised the debate and pressure on greening the economy and fostering a sustainable future that is free from heavy reliance on fossil fuels (Flanagan, 2023) which has been synonymous with global development since the end of World War II.

Higher education institutions are expected to play a central role in this regard and play a leading role in the achievement of not only Sustainable Development Goals (SDG) (Dube, 2024b; Stanic′ & Peterka, 2024; Zhou et al., 2020) under SDG 13 on climate change but also be actively involved in driving the energy transition process. Fossil energy production is often considered a critical driver of GHG accumulation (Lamb et al., 2021; Liu et al., 2019). Addressing the challenges associated with fossil fuel dependency is at the core of ensuring a greener future for all. From an SDG perspective, it is dealt with under SDG 7 which calls for the production of affordable and cleaner energy. Achieving this requires a strong rollout of appropriate technology, a cultural shift and research and development backed by strong advocacy (Aravamudhan et al., 2023; Salkuti, 2022; Song et al., 2024). In all these areas, higher education institutions are expected to play a leading role (Zhang et al., 2023). Higher education institutions have the human and technical capacity to assist in driving this global agenda (Talan & Tyagi, 2020).

Education and higher education institutions play a central role in the development of societies worldwide (Austin & Jones, 2024; King et al., 2011). One of the most celebrated roles is their capability to assist in shaping better societies by championing research, innovation and commercialisation (Atuhaire et al., 2022). Education and higher education institutions are also known for leading policy transformation and development (Ansell, 2015). Private companies and civic society often turn to universities for solutions in an ever-evolving, complex and challenging world. This role is particularly important in society as it relates to the university's role in championing the Sustainable Development Goals (SDGs).

Given the growing centrality of green technological evolution, there is a need to take stock of the role that higher education can play in ensuring sustainability. This study aimed to examine the role of higher education in the green technology revolution. The central question is: What is the role of higher education institutions in the global green technology revolution? This study also seeks to address the challenges faced by higher education institutions in fostering the green technological revolution.

10.2 Methodology

This study utilised data gathered from previous publications Indexed on Web of Science. The search terms used were 'higher education green technology'. The initial search yielded 733 publications published between 2003 and January 2026. The focus was on analysing studies published after the adoption of the Sustainable Development Goals in 2015 and covering a full period. Consequently, after a date restriction was imposed, only 604 documents remained for analysis. These publications were published between 2016 and 2025. Given that the researcher could only grasp English content, a language restriction was placed, resulting in the filtering of 588 articles. Given that the researcher needed a rapid understanding and grasp of the matter at hand, the articles were filtered based on relevance. The top 100 most relevant articles were selected for in-depth analysis. The 100 articles were scrutinised by a thorough reading of topics and abstracts to ensure that all the articles selected could answer the research questions and sub-research questions. After this process, only 57 articles remained for analysis. Following this process, all the remaining articles were extensively read in their entirety for content and meaning. During the process, notes were made on content that answered the research questions.

10.3 Results and Discussions

The study found that higher education institutions play a central role in the green technology revolution, given their inherent capabilities. The role of higher education spans across the three fundamental facets of green technology evolution, namely advancing and fostering the green technological revolution under SDG 9 (Industry Innovation and Infrastructure), SDG 7 (ensuring the rollout of affordable and clean energy) and fostering climate change action from a climate change mitigation perspective under SDG 13 (climate change action). The study also found that higher education institutions can work within the dictates of SDG 4 on quality education in the following ways:

- Fostering Technological Innovation.
- Developing Green Human Capital.
- Promoting Sustainable Practices.
- Encouraging Green Education Policies.
- Facilitating Collaboration and Knowledge Transfer.

10.3.1 The Role of Higher Education in Fostering Green Revolution Under SDG4

Universities and technical colleges play a central role in advancing the ethos set out in the delivery of a quality curriculum. Delivering a quality curriculum from a green technology revolution perspective means that higher education institutions can implement measures aimed at developing a curriculum that is aligned with sustainability and, in particular, aimed at advancing knowledge on greening the planet (Louw, 2013; Ni et al., 2024; Xiong et al., 2013). Higher education institutions (HEIs) play a critical role in delivering curricula that support cognitive skills that promote technological innovation (Zhou et al., 2024).

The study found that higher education institutions can also play a leading role in creating a conducive environment which fosters research and innovation in the production of technological innovation and startup hubs for green technology initiatives (Bouzidi et al., 2024). In other initiatives, HEIs have been central in hackathons aimed at finding solutions to the challenges faced in ensuring green technological development. Such initiatives force the new generation of scholars to be practically engaged in green revolution development. Green innovative initiatives can trickle down to industries and other places where graduates work and assist in steering the world on a greener path (Yusof et al., 2023). Where these kinds of initiatives involve industry partnerships, the rewards have been immense, as they have assisted in ensuring quality education and in reducing the costs of green transitions. Technology transfer has also been accelerated in areas where such arrangements involve the North–South divide. One example of such advances is in the solar energy space, where the cost of solar technology has been decreasing annually, promoting the uptake of such technology.

10.3.2 Higher Education and Green Technological Transfer Offices

The entire ecosystem of green jobs has been driven by advances in the relationship between curriculum and research and innovation at higher education institutions and technological hubs across the world. Initiatives such as green campus initiatives across the world (Shange et al., 2025; Yanthi et al., 2019) have been at the core of creating a sustainability culture among students and other university stakeholders. In Brazil, for example, a collaborative effort between universities has been critical in promoting green technological transfers utilising Technological Transfer Offices (TTOs) (Silva et al., 2023).

This approach allowed the leveraging of several resources which universities and their TTOs can offer, such as a dedicated team of researchers, innovators, commercialisation officers and other ample space for industry collaboration. Consequently, this approach allows universities to play a more meaningful role in bringing together policymakers, the private sector, innovators and businesses. Simultaneously, the approach allowed researchers and developers to access market researchers on challenges and opportunities to evaluate and provide feedback for the improvement of green innovations. Other role players, such as banks, have been seen to be critical in funding such projects through the payment of scholarships to universities for students involved in green innovative projects (Ibid).

According to Santos Silva et al. (2023), the use of TTOs by universities to foster green innovation offers several benefits, as it can enhance teaching and learning. In addition, universities have opportunities to generate additional revenue through licensing and commercialisation of these green technologies. Universities also have opportunities to participate in the development of new products, are better positioned to foster collaboration and contribute meaningfully to economic growth. The participation of universities in the green technology space also allows them better access to funding and expands their research capabilities through expanded opportunities for knowledge exchange. Universities in the space also are able to boost their image, reputation and visibility.

10.3.3 Research and Development in Green Innovation Space

Research and development are another area where higher education institutions have excelled. A number of technological innovations that are currently on the market have been refined and produced by higher education institutions or worked in collaboration with other key players to make them available (Li et al., 2024; Pujotomo et al., 2023; Yi & Zhang, 2022). Research institutions can leverage funding to promote research on green technology development (Bouzidi et al., 2024; da Rosa, 2020). This raises awareness of the need for green technology and encourages green investment. The contributions of HEIs are shown in Fig. 10.1. According to Shafait and Huang (2024), there is considerable value in having leadership in higher education institutions with knowledge of green innovation, as this advances learning outcomes which promotes green innovation and green organizational performance.

The study found that higher education institutions are central to green policy development and often play a central role in policy advocacy (Benhamed & Odeh, 2025), development and implementation. Universities across the world play a critical advocacy role in championing sustainability initiatives and often act as consultancies for various green projects. Higher education institutions also act as cogs to ensure policy alignment, technology transfer and a just energy transition. In some of these projects, universities act as collaborative partners with industry, government and civil society (Deirmentzoglou et al., 2025; Li et al., 2025). In many respects, higher education institutions act as change agents who achieve their goals by transforming campuses (Zaidan et al., 2023) by leading by example in a manner that ensures sustainable operation by targeting and promoting a cultural shift that is aligned to corporate culture.

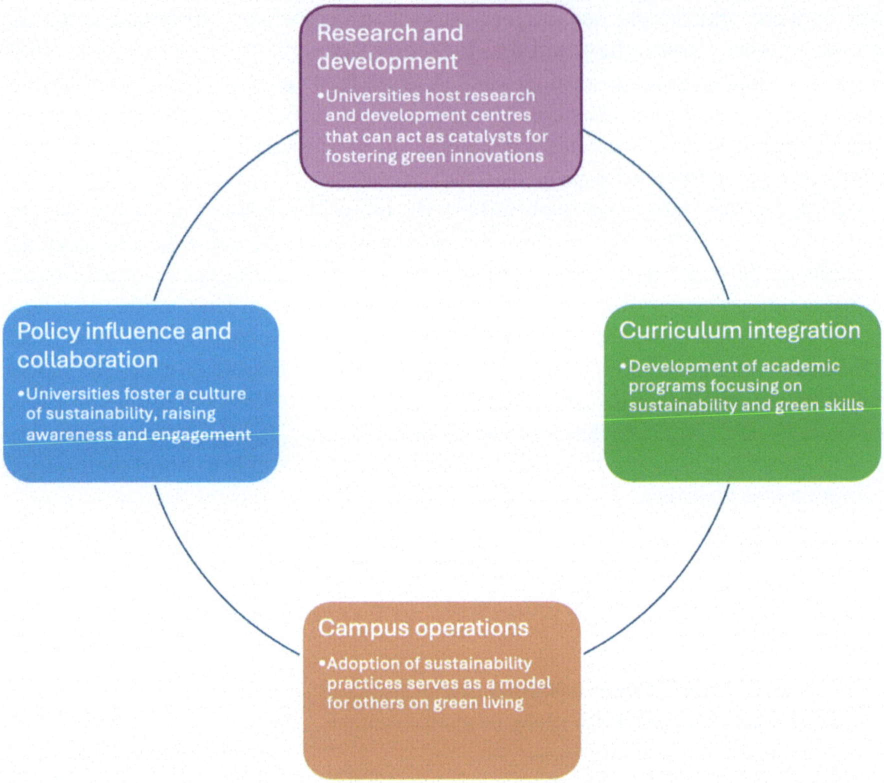

Fig. 10.1 Areas of focus for higher education institutions to contribute to green technology

Apart from the above-stated universities, they leverage their asset endowments, such as campus labs, to test some of the technologies. Through close alignment and collaboration with external partners, real-world validation of some of their developments is ensured, allowing for proper development. Such partnerships can also be extended to support startups and ensure technological transfer (Giones et al., 2022) to provide market-ready solutions in the green space.

10.3.4 Challenged Faced by Higher Education Institutions in Championing Green Technologies

While universities play a central role in fostering green technology, the sector faces numerous challenges that must be addressed to be successful in its endeavour. One of the biggest challenges that higher education institutions face is the lack of technical understanding of the sustainability space and qualified staff who can assist students.

This challenge is particularly visible in developing countries, where green technological transfer and exposure are slow. This means that the level of exposure to some of the latest technology is constrained which hampers efforts at knowledge transfer to students and society.

The other challenge is the disparities in sustainability integration across regions and institutions is the variability in sustainability integration in various regions and also institutions. Other universities also cite disparities in higher education quality which affect innovation outcomes (Sun, 2025). The relationship between universities and industries is often complex, with its own set of challenges. Industry collaboration often presents various challenges, including conflicting interests and mandates.

Equipping some laboratories and setting up infrastructure for green technology to foster the curriculum can be expensive for universities which hampers the speed and pace of joint projects. In some universities, outdated infrastructure causes inefficiencies. Some outdated infrastructure is difficult to deal with as it can be challenging to replace it due to technical complexities.

In some instances, a lack of strong leadership and institutional commitment can also hinder sustainability projects. In the absence of a champion or a specific department which has a passion for greening efforts, initiatives often fail to take root which often results in ad hoc rather than strategic adoption of projects. Some higher-education institutions also lack interest and understanding, resulting in behavioural resistance among faculty members. Some institutions also suffer from bureaucratic challenges which often frustrate inertia towards sustainability projects.

There are also clear challenges associated with curriculum limitations in this study. Updating the curriculum in some settings can be a daunting task which limits progress.

10.4 Conclusions

This study sought to examine the role of higher education in green technology evolution and identify some of the challenges faced by institutions in achieving this goal. The study found that as the quest for green technology and demand grows, higher education institutions can unlock a lot of value that can assist in fostering sustainability. Amongst their duties, higher education has a central role in developing an appropriate curriculum that is responsive to societal needs. Higher education institutions also play a central role in championing green technology evolution through research and innovation. Higher education institutions also play a central role in upscaling green technology and inspiring startups in the green space, among others.

Although the higher education sector has a role to play in fostering green technology to address some of the challenges posed by climate change and environmental degradation, the sector faces numerous challenges in this regard. One of the central challenges is the lack of finances for financing some aspects of green technology evolution. Although there is a desire to ensure sustainability, the lack of green champion leadership can be problematic in some institutions. The speed at which the curriculum can be developed and revised can be problematic.

Higher education institutions need to work closely with accrediting institutions to address some of the challenges faced in curriculum accreditation and review. Given the centrality of green skills, there is a need to ensure that higher education institutions find ways of funding and promoting a green culture in their institutions as a critical component of curriculum. This can be done by unlocking private partnership funding to address financial bottlenecks.

References

Ansell, N. (2015). Shaping global education: International agendas and governmental power. *International Development Planning Review, 37*, 7–16. https://doi.org/10.3828/idpr.2015.2

Aravamudhan, V., Sengodan, A., & Raj, P. M. (2023). Sustainability in conjunction with the latest business technology advancements. In S. G. Yaseen (Ed.), Cutting-edge business technologies in the Big Data era (pp. 47–55). Springer Nature Switzerland.

Atuhaire, S., Higenyi, F., Bacwayo, E., Mugisha, A., Mwebesa, C., & Nambuya, R. (2022). The changing role of higher education institutions on research and innovation in Africa: A systematic review. *International Journal for Innovation Education and Research, 10*, 189–202. https://doi.org/10.31686/ijier.vol10.iss2.3664

Austin, I., & Jones, G. A. (2024). *Governance of higher education: Global perspectives, theories, and practices* (2nd ed.). Routledge.

Badr, A., & El-Shazly, H. H. (2024). Climate change and biodiversity loss: Interconnected challenges and priority measures. *Catrina: The International Journal of Environmental Sciences*. https://doi.org/10.21608/cat.2024.340596

Benhamed, A., & Odeh, M. (2025). Advancing sustainability in higher education institutions: A case study of the higher colleges of technology, Al Ain campuses, UAE, Dr Anji Ben Hamed, Dr Maen Odeh—World Association for Sustainable Development (WASD). https://wasdlibrary.org/download/wjstsd-v20-n1-2-2025-sustainability-he-uae/. Accessed January 18, 2026.

Bouzidi, H., Nadjiba, S., & Khaled, R. (2024). Sustain cubator "Eco Innovate Hub": Nurturing green entrepreneurship in higher education. *International Journal of Economic Perspectives, 18*, 1981–1998.

Chen, X., Li, X., Yuan, X., Zeng, G., Liang, J., Li, X., Xu, W., Luo, Y., & Chen, G. (2018). Effects of human activities and climate change on the reduction of visibility in Beijing over the past 36 years. *Environment International, 116*, 92–100. https://doi.org/10.1016/j.envint.2018.04.009

Chikodzi, D., Dube, K., & Ngcobo, N. (2021). Rethinking harbours, beaches and urban estuaries waste management under climate-induced floods in South Africa. In G. Nhamo & L. Chapungu (Eds.), *The increasing risk of floods and tornadoes in Southern Africa* (pp. 127–140). Springer International Publishing.

da Rosa, M. R. (2020) Adoption of Green IT in the university environment: Systematic review of sustainability practices in educational institutions. *A to Z: Novas Práticas em Informação e Conhecimento, 9*, 79–87. https://doi.org/10.5380/atoz.v9i2.74598

Deirmentzoglou, G. A., Apostolopoulos, N., Apostolopoulos, S., Anastasopoulou, E. E., Topaloglou, L., Nikolaidou, K., Penkova, T., Santamaría, M. C., Nieto-González, S., Jocic, D. R., Stanojević, M., & Sklias, G. (2025). Exploring synergies among European universities, government, industry, and civil society on promotion of green policies and just transition facets: Empirical evidence from six European countries. *Sustainability, 17*, 7517. https://doi.org/10.3390/su17167517

Dong, Y., Zhang, Y., & Liu, S. (2024). The impacts and instruments of energy transition regulations on environmental pollution. *Environmental Impact Assessment Review, 105*, Article 107448. https://doi.org/10.1016/j.eiar.2024.107448

Dube, K. (2024a). El Niño's implications for the Victoria falls resort and tourism economy in the era of climate change. *Sustainability, 16*, 5087. https://doi.org/10.3390/su16125087

Dube, K. (2024b). Challenges and opportunities facing higher education in South Africa. In K. Dube (Ed.), *Redefining education and development: Innovative approaches in the era of the sustainable development goals* (pp. 11–27). Springer Nature Switzerland.

Dube, K., Ezeuduji, I. O., & Swart, M. P. (Eds.) (2025). *Climate change and sustainable development goals discourse in rural and tourism-protected areas*. Springer Nature Switzerland.

Filonchyk, M., Peterson, M. P., Zhang, L., Hurynovich, V., & He, Y. (2024). Greenhouse gases emissions and global climate change: Examining the influence of CO_2, CH_4, and N_2O. *Science of the Total Environment, 935*, Article 173359. https://doi.org/10.1016/j.scitotenv.2024.173359

Flanagan, F. (2023). Just cessation: How might the climate imperative to phase out fossil fuel extraction reshape the concept of just transition? *International Journal of Comparative Labour Law and Industrial Relations, 39*.

Giones, F., Kleine, K., & Tegtmeier, S. (2022). Students as scientists' co-pilots at the onset of technology transfer: A two-way learning process. *The Journal of Technology Transfer, 47*, 1373–1394. https://doi.org/10.1007/s10961-021-09880-3

King, R., Marginson, S., & Naidoo, R. (2011). *Handbook on globalization and higher education*. Edward Elgar Publishing.

Lamb, W. F., Wiedmann, T., Pongratz, J., Andrew, R., Crippa, M., Olivier, J. G. J., Wiedenhofer, D., Mattioli, G., Khourdajie, A. A., House, J., Pachauri, S., Figueroa, M., Saheb, Y., Slade, R., Hubacek, K., Sun, L., Ribeiro, S. K., Khennas, S., de la Rue du Can, S., … Minx, J. (2021). A review of trends and drivers of greenhouse gas emissions by sector from 1990 to 2018. *Environmental Research Letters, 16*, 073005. https://doi.org/10.1088/1748-9326/abee4e

Li, C., Zhang, X., & Wang, L. (2025). How industry–university–research integration promotes green technology innovation in Chinese enterprises: The dual mediating pathways and nonlinear effects. *Systems, 13*, 696. https://doi.org/10.3390/systems13080696

Li, Y., Zhenhong, X., Fu, Z., Wu, Y., & Wang, Y. (2024). Green technology innovation models, environmental regulation and the willingness of government-university-industry to collaborate on green innovation. *Energy Reports, 12*, 2901–2912. https://doi.org/10.1016/j.egyr.2024.08.031

Liu, D., Guo, X., & Xiao, B. (2019). What causes growth of global greenhouse gas emissions? Evidence from 40 countries. *Science of the Total Environment, 661*, 750–766. https://doi.org/10.1016/j.scitotenv.2019.01.197

Louw, W. P. (2013). Green curriculum: Sustainable learning in higher education. *The International Review of Research in Open and Distributed Learning, 14*, 1–15. https://doi.org/10.19173/irrodl.v14i1.1310

Mechler, R., Singh, C., Ebi, K., Djalante, R., Thomas, A., James, R., Tschakert, P., Wewerinke-Singh, M., Schinko, T., Ley, D., Nalau, J., Bouwer, L. M., Huggel, C., Huq, S., Linnerooth-Bayer, J., Surminski, S., Pinho, P., Jones, R., Boyd, E., & Revi, A. (2020). Loss and damage and limits to adaptation: Recent IPCC insights and implications for climate science and policy. *Sustainability Science, 15*, 1245–1251. https://doi.org/10.1007/s11625-020-00807-9

Nguyen, X. P., Hoang, A. T., Ölçer, A. I., & Huynh, T. T. (2025). Record decline in global CO_2 emissions prompted by COVID-19 pandemic and its implications on future climate change policies. *Energy Sources, Part a: Recovery, Utilization, and Environmental Effects, 47*, 4699–4702. https://doi.org/10.1080/15567036.2021.1879969

Ni, L., Ahmad, S. F., Alsanie, G., Lan, N., Irshad, M., Saeed, R. H. B., Ahmad, A. B., & Khan, Y. (2024). Investigating the role of green curriculum in shaping pro-environmental behaviors and environmental values orientation for sustainability. *International Journal of Sustainability in Higher Education, 25*, 1537–1557. https://doi.org/10.1108/IJSHE-05-2023-0207

Pujotomo, D., Hassan, S. A. H. S., Ma'aram, A., & Sutopo, W. (2023) University–industry collaboration in the technology development and technology commercialization stage: A systematic literature review. *Journal of Applied Research in Higher Education, 15*, 1276–1306. https://doi.org/10.1108/JARHE-11-2022-0344

Ray, R. L., Singh, V. P., Singh, S. K., Acharya, B. S., & He, Y. (2022). What is the impact of COVID-19 pandemic on global carbon emissions? *Science of the Total Environment, 816*, Article 151503. https://doi.org/10.1016/j.scitotenv.2021.151503

Salkuti, S. R. (2022). Emerging and advanced green energy technologies for sustainable and resilient future grid. *Energies, 15*, 6667. https://doi.org/10.3390/en15186667

Santos Silva, L. C., Ten Caten, C. S., Gaia, S., & de Oliveira, S. R. (2023). Tool for assessment of the green technology transfer structure in Brazilian public universities. *Sustainability, 15*, 6873. https://doi.org/10.3390/su15086873

Shafait, Z., & Huang, J. (2024). Examining the impact of sustainable leadership on green knowledge sharing and green learning: Understanding the roles of green innovation and green organisational performance. *Journal of Cleaner Production, 457*, Article 142402. https://doi.org/10.1016/j.jclepro.2024.142402

Shange, H. S., Zogli, L.-K.J., & Dlamini, B. I. (2025). Green campus initiatives and strategies for sustainability in higher education. *Transformation in Higher Education, 10*, 364. https://doi.org/10.4102/the.v10i0.364

Silva, L. C. S., Ten Caten, C. S., & Gaia, S. (2023). Conceptual framework of green technology transfer at public university scope Brazilian. *Innovation and Green Development, 2*, Article 100076. https://doi.org/10.1016/j.igd.2023.100076

Song, A., Rasool, Z., Nazar, R., & Anser, M. K. (2024). Towards a greener future: How green technology innovation and energy efficiency are transforming sustainability. *Energy, 290*, Article 129891. https://doi.org/10.1016/j.energy.2023.129891

Stanic′, M., & Peterka, S. O. (2024). Integrating UN sustainable development goals into higher education study programs: Implications and opportunities. In T. Žiljak, P. Ristord, N. Alfirević & J. Pavičić (Eds.), *Lifelong learning for green skills and sustainable development: Southern European perspectives* (pp. 87–100). Springer Nature Switzerland.

Sun, Y. (2025). Mechanisms and pathways of higher education's role in green technology innovation. *Environment, Development and Sustainability*. https://doi.org/10.1007/s10668-025-06566-w

Talan, A., Tyagi, R. D. (2020). Education and human resource development for sustainability. *Sustainability*, 413–438.

Xiong, H., Fu, D., Duan, C., Liu, C., Yang, X., & Wang, R. (2013). Current status of green curriculum in higher education of Mainland China. *Journal of Cleaner Production, 61*, 100–105. https://doi.org/10.1016/j.jclepro.2013.06.033

Yanthi, N., Yunansah, H., Wahyuningsih, Y., & Milama, B. (2019). *Green campus initiative (Where do we start?)* (pp. 39–44). Atlantis Press.

Yi, H., & Zhang, Q. (2022). Knowledge-sharing strategies of university-industry alliances promoting green technology innovation in ecosystems: Based on the utility of multichannel funding. *IEEE Access, 10*, 65728–65743. https://doi.org/10.1109/ACCESS.2022.3184422

Yusof, N. S., Ismail, W. N. A. T., N, N. A. (2023). Factors that influence the practice of green technology among undergraduate students at Universiti Sultan Zainal Abidin. *International Journal of Academic Research in Business and Social Sciences, 13*, 1–8.

Zaidan, E., Belkhiria, E., & Wazen, C. (2023). Universities of the future as catalysts for change: Using the sustainable development goals to reframe sustainability—Qatar university as a case study. *The sustainable university of the future* (pp. 1–23). Springer.

Zhang, T., Ma, Z., & Shang, Y. (2023). Higher education, technological innovation, and green development—Analysis based on China's provincial panel data. *Sustainability, 15*, 4311. https://doi.org/10.3390/su15054311

Zhou, L., Rudhumbu, N., Shumba, J., & Olumide, A. (2020). Role of higher education institutions in the implementation of sustainable development goals. In G. Nhamo & V. Mjimba (Eds.), *Sustainable development goals and institutions of higher education* (pp. 87–96). Springer International Publishing.

Zhou, X., Hu, X., Duan, M., Peng, L., & Zhao, X. (2024). Go for economic transformation and development in China: Financial development, higher education, and green technology evolution. *Evaluation Review, 48*, 32–62. https://doi.org/10.1177/0193841X231166741

Chapter 11
A Practice-Based Exploration of Camera Obscura and Smartphone Photography as a Sustainable and Economical Artistic Practice

Sean Bogle, Mashaole Jacob Makwela, Corne Venter, and Dinkwenyana Malcom Phafane

Abstract This study is inspired by a first-year photography project that focuses on the camera obscura combined with smartphone technology to learn basic photography skills. The exploration of camera obscura and smartphone technology in the photographic process holds the possibility of bridging traditional and automated techniques to generate innovative, economical, and expressive photographs within different thematic contexts. Therefore, in this qualitative practice-based research paper, we argue, through an interpretivist lens, that this basic economical technique has the potential to generate conceptual techniques and sustainable photographic techniques for artists and students with limited or no financial means. The primary data collection method is the production of a body of work by a cohort of four lecturers in a creative group dynamic. The participants answered a project brief with the theme "landscape memory" to explore the potential of camera obscura and smartphone photography. Documentation of the production process includes reflection-in-action and reflection on action in terms of field notes, diaries, and journals. The double helix of the praxis-exergies model was used to interpret the findings. This project and its outcomes aspire to transfer a sustainable and economical artistic practice through a novel photographic technique for artists and students.

S. Bogle · M. J. Makwela (✉) · C. Venter · D. M. Phafane
Department of Visual Arts and Design, Vaal University of Technology, Vanderbijlpark, South Africa
e-mail: mashaole@vut.ac.za

S. Bogle
e-mail: sbogle@vut.ac.za

C. Venter
e-mail: cornev@vut.ac.za

D. M. Phafane
e-mail: dinkwenyanap@vut.ac.za

K. Dube and L. M. Tjabane (eds.), *Advancing Sustainable Development Goals in Higher Education*, https://doi.org/10.1007/978-3-032-27646-9_11

Keywords Camera obscura · Smartphone technology · Economical artistic practice · Landscape memory · Practice-based research

11.1 Introduction

This study was inspired by a first-year photography project. The project focuses on the use of a camera obscura combined with smartphone technology to learn basic photography skills (Bogle, 2024). Based on our teaching and learning experiences in photography courses and modules in fine art and graphic design, the cost of photography technology can be expensive and unsustainable. Khoza and Tshuma (2022) explore a first-year students' digital pinhole photography lesson as a 'mushfake' in resource-constrained environments, such as those with limited funding in higher education. However, a basic online Google search shows that the price of a professional camera body ranges from twenty thousand rands to more than a hundred thousand rands in South Africa. In this context, it is difficult for first-year students or artists to access professional cameras to learn basic photography skills owing to resource constraints.

However, the advent and accessibility of smartphone technology have significantly interrupted traditional approaches to commercial, photojournalism, and fine art photography (Germen, 2014; Piper-Wright, 2020; Strauss et al., 2011). Strauss et al. (2011) discuss the evolution of images and their role in conflicts, highlighting how new technologies and social media have transformed photography. This transformation is evident in the processes of image production, distribution and consumption. According to Germen (2014), the increased use of smartphone photography has provided opportunities for individuals to create and distribute visual stories independently. Piper-Wright (2020) notes that technological advancements in digital photography, such as automation and image preview functions, have significantly reduced common errors like overexposure, blur, and improper cropping. Similarly, Tom and Kearns (n.d.) describe smartphones as portable, convenient, and easy to operate anywhere and anytime. The influence of smartphone photography technology can be observed throughout the entire life cycle of an image, from production to consumption, making it more accessible and cost-effective.

Equally, the camera obscura, as an ancient optical technology, has been integrated with many other technologies used to view transient scenes in art practices, astronomy, and other fields (De Tomasi, 2023; Kiang, 2001). Conventionally, the camera obscura was employed as an instrument by artists for tracing and studying elements of art, such as light, colour, shade, and so forth, in landscape painting (Carson & Shafer, 2008). De Tomasi (2023) notes that the revival of the camera obscura from obsolescence in contemporary artistic practice is reconfigured through the production of images, immersive installations, among other methods. Modern camera obscuras have been developed using 3D printing technology and utilise smartphones to capture and store images (Dietz, 2023). Camera obscura technology is old but can still be repurposed to serve various functions and can be constructed from different materials.

Subsequently, both camera obscura and smartphone technologies are primarily technical devices rather than conceptual tools in the creative photographic process. Langford (2000) reminds us that the photographic process involves 'craft skills, creative decisions, and the intended meaning of the photograph'. This indicates that photographic techniques are not employed as ends in themselves; instead, conceptual and aesthetic choices are equally important. The principle of power within visual culture education also emphasises the importance of understanding the contextual factors that shape the production and reception of visual imagery (Duncum, 2002). This awareness can guide artists in crafting a strong, contextually grounded thematic framework to drive their creativity. Beyond creating captivating visuals, a theme introduces a useful set of parameters for an artist to focus on their creativity and infuse their work with meaningful personal and cultural contexts. Therefore, in this study, we adopted the theme of landscape memory to frame the creative process and explore the use of a camera obscura and smartphone within the domain of landscape photography.

Against this background, we argue that the exploration of the camera obscura and smartphone technology in photographic practice offers the potential to generate sustainable, economical, and expressive techniques within specific thematic contexts. In this study, the concept of sustainable and economical artistic photography practices can be understood in terms of meeting conceptual photographic needs and the affordability of tools, such as smartphone photography and the use of recycled materials. This aligns with the United Nations Sustainable Development Goal 12 (SDG 12), which promotes sustainable consumption and production patterns (United Nations, 2015). The study is guided by the following question: how can the use of camera obscura and smartphone technology in photographic processes generate conceptually rich and economically sustainable techniques? The aim was to explore photographic techniques with aesthetic qualities that are conceptually rich within the context of landscape memory. This was done to ascertain the creative potential and affordability of these mediums, making them suitable for artwork production by students and artists. A practice-based research approach was employed to collect data through the creation of a body of work by a group of four lecturers working in collaboration. The production process was documented through field notes, diaries, and journals. The double helix of the praxis-exergies model (Marshall, 2010) was used to interpret the findings. The results and insights are presented through a visual analysis and reflection on the images produced by the researchers.

11.2 Literature Review

This section engages with key concepts, such as camera obscura, smartphone photography, landscape memory, and sustainable artistic practice. This review aims to understand the possibilities and potential of camera obscura and smartphone technology in the process of creating photographic works Second, it seeks to locate the concept of landscape memory as a theme within the photographic process and to explain the meaning of sustainable artistic practice in the production of photographic work.

11.2.1 Camera Obscura and Smartphone Photography

Camera obscura and smartphone camera technologies are two different devices with distinct methods of image creation. Camera obscura is an optical instrument that uses light to project an inverted image (Kiang, 2001). This device can be described as a dark chamber with a small hole or lens on one side, through which light passes from the outside to project an image onto its interior surface (Keener, 2020). In contrast, a smartphone is a multifunctional device that offers various services, such as camera and communication (Delbracio et al., 2021). Camera obscura is an analogue device that optically projects images, whereas a smartphone camera is a digital tool with multiple functionalities.

Historically, the concept of a dark chamber or dark room combined with light to produce images can be traced back to approximately 400 BC (Taggart & Cole, 2022). The idea of using a dark chamber or dark room with light to generate images alludes to the existence of different types of camera obscuras. These types range from large rooms and portable tents to smaller and more manageable box cameras. In particular, the box camera obscura is small and portable, making it convenient for various applications. Different materials can be used to construct the box camera obscura, such as 3D printing (Dietz, 2023) or cardboard (Bogle, 2024). The box camera obscura can be pointed in any direction, allowing the operator to explore and view the scene in front of them from different angles and at varying distances between the camera and the subject.

Cellphone technology is believed to have been first used and developed on the 3rd of April 1973 (Korn, 2023). In later years, the development of additional services on cellphone technology, such as embedded digital cameras in devices, led to the emergence of smartphones. This development can be linked to digital cameras. In the late 1980s, Fuji Film developed the first commercial handheld digital camera, called the Fuji DS-1P. The images were recorded as 'computerized files' on a 16 MB SRAM internal memory card (Trenholm, 2021). The first instance of a digital camera and phone used to share pictures instantly or in real time on a public network occurred on the 11th of June 1997 (Benezra & Epstein, 2018). Philippe Kahn 'jerry-rigged' a mobile phone with a digital camera and sent photos in real time of the birth of his daughter. What Kahn achieved was the integration of a digital camera and a mobile phone, two separate entities, to perform the function of sharing digital pictures on a public network in real time. According to Germen (2014), the advantages of smartphones include improved communication, portability, connectivity, instant photo sharing, a larger scope of view, and software applications for editing and enhancing images digitally through filters. The potential of camera obscura and smartphone technology in creating photographic works can be viewed in terms of projecting images onto specific surfaces and digitally producing images. Smartphone technology has significantly disrupted traditional approaches to commercial, photojournalism, and fine art photography (Germen, 2014; Piper-Wright, 2020; Strauss et al., 2011). This disruption affects image production, distribution, and consumption (Strauss et al., 2011). The image creation process is influenced not only by the technical capabilities of devices but also by conceptual visual representations. According to De Tomasi (2023), the camera obscura in contemporary art is reconfigured through image production and immersive installations. Similarly, Delbracio et al. (2021) observed that computational photography technology for mobile devices is designed to produce aesthetically pleasing images, especially through post-processing and manipulation to enhance quality and aesthetics. Therefore, smartphone photography capabilities can serve as a conceptual tool within the creative photographic process, expanding artistic possibilities.

11.2.2 The Concept of Landscape Memory in a Photographic Process

Section 11.2.1 demonstrates the characteristics of the camera obscura and the technical capabilities of smartphone cameras in capturing photographs. In this section, it is important to locate the concept of landscape memory as a theme within the domain of landscape photography and how the visual properties of camera obscura and smartphone photography are interrelated with the concept of memory. According to Langford (2000), successful photographs have meaning. To understand and create such meanings in photographs, contextual factors shape the production and reception of visual imagery (Duncum, 2002). In contrast, the concept of 'landscape', transcends the traditional notion of a mere geographic terrain, encompassing the multifaceted and dynamic interactions between people, places, and time. Schama (1995) highlights that 'even the landscapes that we are supposed to consider most free of our culture may turn out, on closer inspection, to be its product'.

Landscape, along with notions of territory, ownership, and colonisation, has been central to South Africa's struggles throughout its difficult past and contentious present (Radebe & Zodwa, 2023). Schama (1995) also notes that landscapes are cultural creations imagined and projected onto rocks, wood, and water before they are natural. However, he states that it must equally be recognised that once a specific concept of landscape—a myth, a vision—takes root in a real location, it blurs the lines between categories, giving metaphors greater life than their physical appearances.

Geng (2024) likens landscape memory to a computer memory that '[i]t is a store of information that can be accessed by a user'. This can be experienced through the act of remembering, encompassing histories of joy, trauma, and everything in between (Marschall, 2013). Personal and cultural memories are often refracted through individual perspectives and collective experiences (Kühn, 2007). Photographic representations of landscapes reproduce natural or man-made environments and reflect the photographer's interpretations, which can be influenced by their experiences or history (De Klerk, 2020; Kirkwood, 2023). Therefore, the theme of landscape memory is suitable for exploring the use of camera obscuras and smartphones at a conceptual level within the domain of landscape photography.

The relationship between photography and memory can be understood from two perspectives. Shobeiri (2023) identifies these as 'conceiving photography through memory' and 'perceiving memory through photography'. These approaches involve creating memories by recording experiences and interpreting them through photographs. Tsakiridou (2021) describes the act of recording experiences as capturing the 'defacto present first, and aesthetically present second', meaning that the subject or object must exist first and then be represented aesthetically. The visual properties of the camera obscura, as used by Abelardo Morell, are also linked to memory. Campbell (2020) characterised Morell's camera obscura photographs as wondrous, dreamlike, and abstract. De Tomasi (2023) considers Morell's work as meta-photography because of its layered interior spaces and projections, which create a visual aesthetic spectacle. These aesthetic qualities, derived from such processes and analyses of other artists' photographs, can inform the development of the concept of memory in the production of artwork.

11.2.3 Sustainable and Economical Artistic Practice

The concept of sustainability can be understood from various perspectives. For example, it involves meeting present and future needs (United Nations Brundtland Commission, 1987), maintaining the capacity to sustain something over time (Basiago, 1999), preserving stable capital (Dresner, 2008), and ensuring that businesses endure over the long term (Giovannoni & Fabietti, 2013). Although sustainability can be interpreted differently depending on the context, there is a common understanding that it involves maintaining something so that it lasts over time. According to the United Nations Environment Programme (2015), social, environmental, and economic factors must be addressed simultaneously to achieve this goal. The United Nations Sustainable Development Goal (SDG) 12 emphasises the importance of ensuring sustainable consumption and production patterns. This implies that the production and consumption of goods and services should not deplete natural resources or harm the environment, thereby safeguarding resources for future generations. Technology and lifestyle choices are considered the primary drivers of this process. For instance, the development of traditional photography technology has shifted from chemical processing in darkrooms to digital photography (Piper-Wright & Jussa, 2024), exemplifying how technological advancements can promote sustainability by reducing environmental impact.

The affordability of materials and tools, such as cardboard box camera obscura and smartphone photography, addresses the consumption and production of services and products. According to Bogle (2024), the cardboard box camera obscura is built from recyclable materials. This makes it a good choice for students learning the basic principles of image formation. By constructing a camera out of a cardboard box, using a magnifying glass as the lens, and translucent material such as wax paper, the cost is minimal. In addition, the translucent material inside the camera can be adjusted to explore and focus on objects before capturing images with a smartphone camera.

In contrast to the room or tent camera obscura, Abelarado Morell (2024) is one of the few modern artists who have incorporated the camera obscura into their practice. He projects images onto surfaces such as the ground, paving, and wooden decks, which become an integral part of the visual representation of the environment. Similarly, the translucent screen on which the image is viewed in the Do It Yourself (DIY) camera project can vary in size, texture, and shape, all of which influence the final presentation of the visual. A smartphone camera is capable of capturing and recording images and videos (Rouse, 2017). Its technology and image data processing are comparable to those of entry-level and mid-range mirrorless digital cameras. Compared to larger digital cameras, smartphone cameras are smaller, portable, and have smaller sensor sizes, and advancements in computational photography (Piper-Wright, 2020; Tom & Kearns, n.d.) have enhanced their capabilities. The use of recycled materials in the construction of camera obscuras and the integration of smartphone technology with multilayered digital camera features demonstrate that these devices can serve as sustainable and cost-effective means of production.

11.2.4 Research Methodology

This research project adopted a qualitative, practice-based approach through an interpretivist lens. According to Leavy (2009), 'art-based research practices are particularly useful for research projects that aim to describe, explore, or discover; methods are generally attentive to processes'. Sullivan (2010) defines art practice as research that 'examines the process of theorizing and uses conceptual problem-solving strategies to analyse and synthesize things in order to explain them in ways that help to implement new practices'. This type of research can be described as inductive in its approach to knowledge building through explorative methods. Therefore, the primary content generated in art-based research involved the production of photographic work using camera obscura and smartphone technology. Ligget (2020) emphasises that artistic creation is fully understood through practice, encompassing not only physical activity but also influences, ideas, and critical reflections, and summarises practice-based research.

Artistic creation within the context of practice-based research is regarded as original, and new knowledge can be acquired through practice and the artwork produced. In this assessment, four art educator researchers in a creative group dynamic responded to a project brief to produce a sample of four photographic artworks, each utilising the camera obscura and smartphone technology in their practice across the fields of fine art, graphic design, and photography. The process was guided by the theme of landscape memory, which allowed for the exploration of the interpretation, sensibility, and interests of the art educator researchers. The creation of photographic artworks involved four phases: (a) literature review, (b) building the camera obscura, (c) producing the artworks, and (d) writing reflections on the results. Phases c and d required participants to collect data through notes, diaries, sketches, photographs, tests, and group meetings, all based on the principles of reflection-in-action and reflection on action.

The objective of the production of the artworks was to utilise the constructed camera obscura and smartphone, focusing on the theme of landscape memory. This approach offers an opportunity for a creative thought process to visualise information presented at the time of creation and explore how one could alter or adapt the photographic work. The artist is also afforded the chance to experiment with 'errors' in the photographic process, as Piper-Wright (2020) mentions. The image may then be manipulated, altered, or the focus points changed, all before recording the visual with a smartphone camera. This process was conducted in situ, and the results of the photographic process were enhanced using smartphone applications or Adobe software. The documentation of the production process includes reflection-in-action and reflection on action, documented through field notes, test photographs, and group meetings. The data analysis strategies used to interpret and reflect on the findings are based on the double helix of the praxis-exegesis model. According to Marshall (2010), the double helix process is reflexive and involves both practice and theory through critical reflection and action. Ligget (2020) provides a useful description of what an artist's practice in the context of research entails, stating that: 'The word "practice" is used in the arts as it is in other professions to recognize knowledge production through action and doing'.

11.3 Presentation and Discussion of Findings

The analysis and findings presented in this paper adopt a three-pronged approach structured into three phases: the construction of the camera obscura, creation of the artworks, and written reflections on the process. This methodology was guided by the research question and the landscape memory theme and sustainable methods employed during the photographic process. Three sites of the image, as coined by Rose (2007) for analysing artworks—namely production, the image itself, and the audience—were utilised through thematic analysis, content analysis, and visual metaphor analysis.

11.3.1 *Building of the Camera Obscura and the Smartphone Models Used*

All four participants constructed their camera obscuras following the same project brief, with different cardboard box sizes but using a similar 60 mm magnifying lens (see Fig. 11.1). The construction of the camera obscura was guided by the following steps: First, a cardboard box was sourced, and an opening was made at the back of the box to accommodate the smartphone used for taking photographs. The different forms, sizes, and shapes of the selected boxes resulted in varying aspect ratios, but the photographs produced exhibited similar effects and textures (see Figs. 11.2, 11.3, 11.4, 11.5, 11.6, 11.7, 11.8, 11.9, 11.10, 11.11, 11.12, 11.13, 11.14, 11.15, 11.16 and 11.17). After this step, a 60-mm lens (taken from a magnifying glass) was mounted at the front of the box; this lens acted as the medium through which the image was reflected.

The final step in the construction process was to create a wireframe onto which a substrate was attached, serving as a screen for the reflected image. This frame was designed to be movable to facilitate focusing the image on the substrate before capturing it with the smartphone. Some participants used waxed baking paper as a focusing instrument, while others experimented with various materials such as tracing paper, fabric, and tissue paper. Ultimately, waxed baking and tissue papers were used for the final fieldwork. The choice of materials was based on the different visual effects they produced on the skin. For example, tissue paper imparts a grainy quality to the images, distorting the visuals and elevating them beyond simple representational photography into a surreal realm. The decision regarding which smartphone model

Fig. 11.1 Cardboard box camera obscura. *Source* Authors

to use for capturing photographs was limited to everyday smartphones for private use, such as the Samsung Galaxy G23, A23, S22, and A54.

Therefore, the construction of the camera obscura can vary in terms of size, shape, and materials used for the focusing instrument, depending on the availability of materials or the interests of the art educator-researcher. The choice of smartphone can range from high-end to basic models, with a minimum resolution of at least 12 megapixels. This ensures that the final images can be printed at a standard or medium resolution suitable for exhibition purposes. These parameters also promote sustainable practices, as no new equipment is required; a camera obscura can be constructed from found objects, emphasising resourcefulness and environmental awareness.

11.3.2 Production of the Art Works and Reflection on Action

This section presents the participants' photographic process through three categories: (a) concept development, focusing on ideas and thoughts during the planning and shooting phases within the context of the landscape of memory theme; (b) the use of economically sustainable techniques to explore the theme; and (c) aesthetic considerations, emphasising elements and principles of design. It is important to note that artistic practice, thinking, and making processes do not occur in a linear or strictly structured manner but in a cyclical process. These processes often involve multiple iterations before arriving at a satisfactory solution (Marshall, 2010).

11.3.2.1 Concept Development

There is no single approach to planning and conceptualising the artistic photographic process; however, there are similar thought processes that influence our experiences and preferences. Five themes emerged in the thinking process during the pre-production of the project concerning the choices of locations or landscapes: personal memory, road trips (visiting family or commuting to and from work), principles of photography, found objects, and African proverbs. Educator-researcher 3, for example, explained his thought process as follows: he has long been fascinated by early photographic pioneers such as Joseph Nicéphore Niépce and Louis Daguerre. Their images attract him with their lengthy exposure times, which often lead to unintended abstraction through the movement of subjects. Their photographs feel more akin to drawings than traditional photographs, and he finds the blurred movements to be visceral captures of the passage of time. Much like later Futurist painters, these early photographs captured the movement of objects and people through time. This medium provides him with the opportunity to capture such abstract moments almost inadvertently. Additionally, the influence of found objects, such as Monobloc chairs—derived from the plastic injection process where one chair can be made from a single (mono) piece of moulded plastic—and often referred to simply as "the plastic or garden chair", is notable in his work. These chairs are available in various colours and designs, yet they share a common function: to be easily stackable, durable, and, most importantly, cost-effective. It is arguably the most mundane and ubiquitous example of the modernist mantra: *form follows function*. In his case, the pink chair can be seen as 'an extension of my own identity, a signifier of gender and sexual norms, or a playful intervention in the natural landscape'.

While educator-researcher 1 explored areas close to home, such as easy walks along paths and open fields, this exploration followed the commute to and from work as a potential area to discover and photograph. In a similar approach, educator-researcher 2 identified possible locations within five kilometres from his home, such as main roads, train stations, and maize fields. The thinking process in this regard is influenced by personal memories. Other locations are influenced by road trips to visit family and friends. This illustrates that landscape photography can be spontaneous and evolve from initial concepts. Using the lens of African proverbs and personal memory of struggle, survival, and acceptance in Sharpeville, educator-researcher 4 interprets university recreational spaces. Furthermore, the township landscape of Sharpeville raises debates about substandard infrastructure and the degradation of social life.

The township milieu serves as a backdrop for diverse people with different dreams and goals to come together. Sharpeville incubates the often fraught lives of its residents according to their own perspectives. Life in these spaces can be a perpetual struggle for survival and acceptance, oscillating between the binaries of good and evil; yet, the in between, that is, the ordinary, persists in the township. Personal memory is shaped by individual lived experiences, where memory and landscape intersect in the process of conceptualisation, influencing the creative decisions made by the photographer. According to Marschall (2013), individuals negotiate their identities, often by revealing the socio-political dimensions of their experiences as they navigate different spaces.

The participants' different thought processes became evident when considering the planning and conceptualisation of the project brief. Personal memories and lived experiences play important roles in deciding shooting locations. However, the decision-making process regarding locations is also influenced by locality or distance to limit travel costs. It is also apparent that they took advantage of road trips—whether for work or family visits—to reduce their expenses.

11.3.2.2 Techniques Used to Explore the Landscape Theme and Aesthetics

Similar to the pre-production creative thought process, the second phase—the actual taking of photographs in the field—requires a combination of theoretical concepts or themes and practical procedures. Jia (2022) emphasises that 'the most important thing of excellent photography is clear theme and accurate conception'. During this stage, decisions must be made regarding what is to be captured and how to best portray such spaces and landscape memories, focusing on techniques and aesthetics. This phase is also reflexive and cyclical in nature; decisions are based on reflection-in-action and reflection on action (Marshall, 2010). This process necessitates that participants reflect on the physical locale to inform their creative and technical decisions and then extend their reflections after the photographs are captured and reviewed offsite. Although the concepts of technique and aesthetic considerations may connect the participants, it is evident that their application varies according to each participant's interests and perceptions. Therefore, the discussion of the visuals produced by each participant will be referred to as explorations.

(i) *Exploration by the Educator-Researcher One*

The image in Fig. 11.2 depicts a colourful urban landscape featuring a tree viewed from a pathway. The camera was tilted upward to capture the branches of the tree. The focus point was centred in the frame, with blurring occurring towards the edges. The texture and blemishes are a direct result of the wax paper used, upon which the image is projected by a magnifying glass lens. This technique created a visual with strong linear and organic lines of the tree branches protruding into the foreground. The branches, both in and out of focus, form subjects that dominate the frame. The space reflected here is part of my daily commute to work. The camera obscura and urban space reflect a world of beauty and charm, a place of memory that echoes both past and present.

The image in Fig. 11.3, using the same camera obscura pointed upward towards the sun and moved close to a spiky-shaped plant, plays with dark tones on the right side of the image against the light tones of the sun in the centre. The simple magnifying lens captured details in both sharp and soft focus simultaneously, resulting in a dreamlike representation of the urban landscape. An interesting aspect of the image is the screen at the bottom left, which receives no light. What you see is the image recorded on the screen inside the camera, but with no light reaching it, resulting in no image formation in that region. This adds an air of mystery to the image's ambiance. I also took pictures in my backyard.

Fig. 11.2 '*The tree of life*'. Digital photography. *Source* Authors

Fig. 11.3 '*Paper worlds*'. Digital photography. *Source* Authors

In Fig. 11.4, *titled Beauty in the Beast*, the image was taken from inside my car, looking outwards. In open spaces with little to no shaded areas, viewing the smartphone camera screen when capturing images projected inside the camera obscura is challenging.

This is due to the ambient light in the environment. To counter this, the viewing screen must be shaded to see clearly. From inside the car, you are shaded and can view the screen at the back of your smartphone camera. The combination of the lens and textured image screen created beautiful textures and recorded fine details of the subject on the right side of the frame. This space is located just off a gravel road on the way to work.

In Fig. 11.5, titled *Fairy Wonderland*, the camera obscura was pushed into a shrub and pointed upward. I then proceeded to move the camera around—left, right, up, and down—changing the camera's angle of view as well as the viewpoint, with the

Fig. 11.4 '*Beauty in the beast*'. Digital photography. *Source* Authors

Fig. 11.5 *'Fairy Wonderland'*. Digital photography. *Source* Authors

images changing as the camera moved. I observed the changes on the wax paper screen as they flashed before my eyes. The movements were slow, as I observed the changing screens. I stopped, and the image on my screen was that of a fantasy world—a beautiful array of colours and shapes. "Fairy Wonderland" is a micro-exploration of a shrub in my backyard. This provided an opportunity to create a better aesthetic image, with magnifications of the scene in front of you being enlarged and seen as a colour abstract of a plant in a garden. The abstraction created a space where, when viewed, one could see a valley, rocks, and shrubbery.

Figures 11.6, 11.7, 11.8 and 11.9 were taken in the evening to take advantage of the golden hour and capture strong-contrast images that could express personal memories. Additionally, the subject matter of the photographs was guided by interesting objects or scenes that resonated with their personal memories. However, other captured photographs were primarily chosen based on their quality and aesthetic appeal rather than personal significance. Approximately 200 photographs were taken over four months. The top ten photographs were selected based on image exposure quality, uniqueness of shot angles, and the overall feel in terms of colour, mood, and tone within the context of memory. Ultimately, only the best four photographs were selected: Fig. 11.6: *Landmark*; Fig. 11.7: *Agave*; Fig. 11.8: *Odour*; and Fig. 11.9: *Shosholoza.*

Figures 11.6, 11.7, 11.8 and 11.9 were taken during a road trip and in a township in East Rand, Gauteng Province. Figure 11.6, titled *Landmark*, depicts the position of an old tree situated in the middle of nowhere, which resonated with my childhood experiences through the mood and tone of the photograph. In my childhood in Limpopo, large and old trees served as landmarks for navigation, such as the Marula tree along the road. These trees can live for more than a hundred years but are vulnerable to lightning strikes. Figure 11.7, titled *Agave (or century plant)*, and Fig. 11.8, titled *Odour*, were also captured during the trip. Additionally, Fig. 11.9, titled *Shosholoza*, was captured within three kilometres of my current residence. The imagery and titles of these photographs reflect both the natural environment and cultural elements of

Fig. 11.6 ‘*Landmark*’. Digital photography. *Source* Authors

the region, emphasising the significance of landmarks and natural features in local navigation and cultural identity of the people.

Similar to Fig. 11.7, *Agave* (or century plant), although the sharp and spiky feel influenced the capturing of the photograph, the blurry focus of the background created a dry environment that can represent winter.

This evokes memories of how we used to create handwork from Agave (or century plant) fibre during primary school. In the context of a road trip, Fig. 11.8 shows that *odour* serves as a metaphor for the smell of sewage in townships, which changes the landscape through scent and furrows in the streets, thereby affecting the lived experiences of visitors and the community at large. This photograph, within the context of landscape memory, comments on how positive memories of the environment can be reshaped daily.

Finally, Fig. 11.9: *Shosholoza* resonates with my personal experiences as a Technikon student in the early 2000s, and today, the space functions as my neighbourhood. The dilapidated train terminals and the silhouettes of electrical power lines in the

Fig. 11.7 *Agave*. Digital photography. *Source* Authors

Fig. 11.8 '*odour*'. Digital photography (photograph by educator-researcher 2)

distance evoke memories of travelling from Vereeniging to Germiston to Tembisa, as it was one of the most affordable modes of transportation.

(ii) *Exploration by the Educator-Researcher Three*

The first photo (Fig. 11.10) was taken outside a church in Swartruggens as I travelled to Groot Marico. In this instance, the scene benefits greatly from the technique of the camera obscura, as brooding clouds hang over the grainy church structure with the bright pink chair in the foreground, becoming a metaphor for my relationship with organised religion and my misgivings about it in general.

The area we were headed to was known to be conservative in its views, and thus the image signifies to me the foreboding and angst that I felt as a gay teen, socialised as Afrikaans, when I entered a church. The fact that I could not enter the church grounds and had to take the photo outside the border fence of its property underscores this feeling of alienation and conditional belonging. I took my next photos (Figs. 11.10,

Fig. 11.9 '*Shosholoza*'. Digital photography. *Source* Authors

Fig. 11.10 '*Swartruggens 1*'. Digital photography. *Source* Authors

Fig. 11.11 '*Martha Glatthaar Monobloc 1*'. Digital photography. *Source* Authors

11.11, 11.12 and 11.13) at the site of the old *Veldskool*[1] I had attended so many years ago. The school has changed a lot since I last saw it, but the environment remains as I recall—hot and humid, verdant bushveld dissected by bright red dust roads.

The school was also behind a locked gate, and again, I found echoes of the exclusion of that time. However, this time, the school and its grounds reminded me of the tenuous association I have always had with accepted notions of masculinity. When I attended that school, South Africa and the world were much different; they were more prescriptive about what was allowed and what was not allowed.

These restrictive notions of gender have eroded over the years, but finding an abandoned building among the overgrowth reminded me of how repressed and stifling it was as an adolescent to be queer in a heteronormative society. Specifically, at this site, where we were taken as a group of boys to live in nature together—for most of us for the first time without our families—and to participate in and compete in '*masculine activities*' such as night marches, obstacle courses, and the dreaded

[1] As van Wyk (2024) describes it: 'Veldskool. If you are white and were in the northern schools under the Transvaal Education Department from the late 1970s to the early 1990s, you will know what it was. I don't really know if Veldskool should be spelt with a capital letter, but it feels that way because it was a much bigger institution than the name suggests.'

Fig. 11.12 '*Martha Glatthaar Monobloc 2*'. Digital photography. *Source* Authors

Fig. 11.13 *Literia 1*. Digital photography. *Source* Authors

Lanternjag. Reflecting on it now, the memory of *Veldskool* is filled with the sense of standing on the edge of adulthood—a feeling as wild, lush, and dangerous as the bushveld itself.

(iii) *Exploration by the Educator-Researcher four*

Figure 11.14, *Dreams of Concrete*, represents the university building. As someone born and raised in the Vaal, this university not only symbolises architectural durability but also serves as a monumental space that helps South Africans from the periphery to realise their dreams. Although the university building was photographed under the scorching summer afternoon light, the results showed minimal and faint illumination. Interestingly, the lighting is more concentrated in the centre than in the corners, creating a moody atmosphere with a strong sense of vignette.

Figure 11.15, *Thinking Out of the Box*, depicts the university's recreational space, parking lot and the surrounding trees. The photograph features a frame within a frame: the outer frame is created by the insides of the camera obscura, which also forms part of the photograph itself. The inner frame is divided into two triangles representing the

Fig. 11.14 '*Dreams of concrete*'. Digital photography. *Source* Authors

sky and landscape. Conceptually and aesthetically, this photographic process creates a layered and meaningful composition.

Figure 11.16 illustrates an intersection joining Barrage Road, next to the university. The image included four human figures walking along the road. South African roads are often crowded with pedestrians. Named after the Northern Sotho proverb,

Fig. 11.15 '*Thinking outside the box*'. Digital photography. *Source* Authors

Fig. 11.16 *'Tshipu e rile ke lebelo, mohlaba wa re ke namile'*. *Source* Authors

Tshipu e rile ke lebelo, mohlaba wa re ke namile, which translates to 'the fox said he is fast, and the landscape said but I am vast', this figure captures the endless encounter between Africans and their landscape, which seems infinite. Africans are deeply infused with their environment, and this image reflects their ongoing relationship.

Figure 11.17, *Choices*, refers to the common perception among many South Africans that if you have seen one township, you have seen them all. This view is often based on substandard infrastructure and social degradation. However, these spaces are home to diverse populations with different aspirations and opportunities. For example, Sharpeville inherently nurtured the fragile lives of its residents, as seen from various perspectives. This photograph benefits from bracketing, a technique that allows capturing the same scene with three different exposures, enabling the selection of the best image.

In summary, although the construction of the obscura camera can differ in size and the smartphone camera specifications used by the participants, the important goal is the potential of inexpensive technology that can generate images of at least 12 megapixels. Second, the participants' artistic concept development processes involved different strategies for planning and idea generation that were influenced by personal memory and lived experiences. The theme of landscape memory contextualises technical planning and thinking in terms of aesthetic expression. Finally, the participants used similar techniques in their production process to explore the landscape theme and aesthetics to depict content about memory, which was influenced by the art educator-researcher's interests. This essentially shows how a personal style can create meaning in photographs. For example, some participants focused more on the principles of photography and aesthetics than on the concepts of meaning-making

Fig. 11.17 '*Choices*'. Digital photography. *Source* Authors

behind creating the artwork. This could mean that the form and content of the generated photographs can create visual narrative memories, offering alternative ways of engaging with the past and present that resonate with diverse lived experiences using obscura cameras and smartphone technology. This approach invites viewers to engage with the complexities of memory in a manner that balances traditional and contemporary technology. In doing so, the camera obscura becomes a powerful medium for enabling a deeper understanding of the intricate relationship between memory and landscape, encouraging reflection not only on the images produced but also on the rich narratives they represent, ultimately enriching our collective journey of self-discovery.

11.4 Conclusion

Camera obscura and smartphone technology in a photographic process suggest that traditional and automated techniques can generate different artistic photographs through conceptual and cost-effective practices. The practice of bridging old and new technologies provides educators, students, and artists with alternative approaches to creating conceptual photographic work. Understanding the technology, economical techniques, and contextual frameworks, such as themes, allows educator researchers to generate compelling, expressive, and aesthetically pleasing photographic works.

The conceptualisation and planning phases of the project are primarily dominated by understanding the content and context. Meanwhile, the production phase is defined by the techniques and aesthetic choices made by the educator researchers.

Therefore, these findings lay the groundwork for educators, students, and artists to develop transferable practices that bridge old and new technologies to create fresh and alternative approaches to creating conceptually rich and innovative visuals. The limitation of this study is that it only focuses on the photographic process to generate conceptually rich, economically sustainable techniques for artists or students, rather than on the reception of the photographic work. Therefore, further research should explore the reception of artworks to determine their success in the context of sustainable economic products. In addition, the application of cost-effective materials to maximise the impact of sustainable development in education can be explored in curriculum modules or workshops for underserved communities.

References

Basiago, A. D. (1999). *Economic, social, and environmental sustainability in development theory and urban planning practice: The environmentalist*. Kluwer Academic Publishers.

Benezra, S., & Epstein, S. S. (2018). Museum of the moving image: The birth of the camera phone. https://scienceandfilm.org/articles/3126/the-birth-of-the-camera-phone. Accessed August 12, 2024.

Bogle, S. (2024). *Department of visual arts and design: Study guide for theory of media technologies 1 (HDTMT1A)*. Vaal University of Technology.

Campbell, H. (2020). Camera in camera. *Places Journal*, October 2020. Accessed January 16, 2026. https://doi.org/10.22269/201006

Carson, J., & Shafer, A. (2008). West, Copley, and the Camera Obscura. *American Art, 22*(2) (Summer 2008), 24–41. https://doi.org/10.1086/591168

De Klerk, A. (2020). Photographer–camera–place relations: Reflections on postphenomenology and landscape photography practice. *Visual Studies, 35*(2–3), 201–215. https://doi.org/10.1080/1472586X.2020.1779608

De Tomasi, F. (2023). Notes on *camera obscura*: Three contemporary artistic perspectives on the path of photography. *Arte, Individuo y Sociedad, 35*(4), 1397–1414. https://doi.org/10.5209/aris.87944

Delbracio, M., Kelly, D., Brown, M. S., & Milanfar, P. (2021). Mobile computational photography: A tour. *Annual Review of Vision Science,* 7, 571–604. https://doi.org/10.1146/annurev-vision-093019-115521

Dietz, H. G. (2023). Digital camera Obscuras. In *IS&T International symposium on electronic imaging 2023, imaging sensors and systems 2023*. https://doi.org/10.2352/EI.2023.35.6.ISS-334.

Duncum, P. (2002). Visual culture art education: Why, what and how. *Wiley, 21*(1), 14–23. https://doi.org/10.1111/1468-5949.00292

Dresner, S. (2008). *The principles of sustainability*. Earthscan.

Geng, Z. (2024). The evolution of landscape photography: From humanistic framing to Posthuman Simulacra. *Costume and Culture Studies, 2*(4), 12–18. https://doi.org/10.6914/ccs.020402

Germen, M. (2014). The unstoppable rise of mobile imaging and aesthetics. *Mobile imaging. Individuality. Digital aesthetics. Visual communication. Contemporary photography*. https://doi.org/10.14236/ewic/eva2014.35

Giovannoni, E., & Fabietti, G. (2013). What is sustainability? A review of the concept and its applications. In C. Busco, M. L. Frigo, A. Riccaboni & P. Quatrone (Eds.), *Integrated reporting, concepts and cases that redefine corporate accountability*. (pp. 21–40). Springer International Publishing.

Jia, Y. (2022). Strategies to train and improve students' composition creation ability in photography practice teaching. *Advances in Social Science, Education and Humanities Research, 635*, 336–341. https://doi.org/10.2991/assehr.k.220107.065.

Keener, K. (2020). Art critique: An illustration of how a basic camera obscura works. Courtesy Wikimedia Commons. https://www.art-critique.com/en/2020/03/a-lesson-on-the-camera-obscura/. Accessed June 9, 2024.

Khoza, B. J., & Tshuma, N. (2022). Photography education in resource-constrained contexts: Exploring the potential of mushfaking. *International Journal of Mobile and Blended Learning, 14*(3).

Kiang, T. (2001). Dark wonder. Under the microscope: Special art and science issue, *Circa, 96*(Summer, 2001), 22–24.

Kirkwood, M. L. E. (2023). Alternative methods in contemporary South African landscape photography. *Photographies, 16*(3), 434–451. https://doi.org/10.1080/17540763.2023.2237966

Korn, J. (2023). 50 years ago, he made the first smartphone call. https://edition.cnn.com/2023/04/03/tech/cell-phone-turns-50/index.html. Accessed August 12, 2024.

Kühn, A. (2007). Photography and cultural memory: A methodological exploration. *Taylor & Francis, 22*(3), 283–292. https://doi.org/10.1080/14725860701657175

Langford, M. (2000). *Basic photography* (7th ed.). Focal Press.

Leavy, P. (2009). *Methods meets art, art-based research practice.* The Guilford Press.

Ligget, S. (2020). Positioning the arts in the research process: Perspectives from higher education. In R. Earnshaw et al. (Eds.), *Technology, design and the arts—Opportunities and challenges. Springer series on cultural computing.* https://doi.org/10.1007/978-3-030-42097-0.9-20

Marschall, S. (2013). Collective memory and cultural difference: Official vs. vernacular forms of commemorating the past. *Taylor & Francis, 14*(1), 77–92. https://doi.org/10.1080/17533171.2012.760832

Marshall, C. (2010). A research design for studio-based research in art. *Teaching Artist Journal, 8*(2), 77–87. https://doi.org/10.1080/15411791003618597

Morell, A. (2024). Camera Obscura: Selected works/camera-obscura. https://www.abelardomorell.net/selectedworks/camera-obscura. Accessed June 9, 2024.

Piper-Wright, T., & Jussa, T. (2024). Quick, quick, slow: making time for sustainable photography practices in contemporary higher education. *International Journal of Art & Design Education, 43*(4), 615–629.

Piper-Wright, T. (2020). Between presence and program: The photographic error as counterculture. In R. Earnshaw et al. (Eds.), *Technology, design and the arts - opportunities and challenges, Springer series on cultural computing.* https://doi.org/10.1007/978-3-030-42097-0.139-157

Radebe, N., & Zodwa, N. (2023). *Singabantu bendawo*: understanding the concept of land from the perspective of Ubuntu. https://doi.org/10.1080/23323256.2019.1670086. Accessed on August 24, 2024.

Rose, G. (2007). *Visual methodologies. An introduction to the interpretation of visual materials.* Sage publications.

Rouse, M. (2017). Techopedia, Camera Phone. https://www.techopedia.com/definition/24224/camera-phone. Accessed August 22, 2024.

Schama, S. (1995). *Landscape and memory.* Random House.

Shobeiri, A. (2024). Photography and memory. In: L. M. Bietti & M. Pogacar (Eds.), *The palgrave encyclopedia of memory studies* (pp. 1–10). Palgrave Macmillan, Cham. https://doi.org/10.1007/978-3-030-93789-8_33-2

Strauss, D. L., Rashidi, Y. E., Sekula, A., Solomon-Godeau, A., Sealy, M., Keenan, T., Egoyan, A., Jaar, A., Willis, D., Cole, D., Azoulay, A., Addario, L., Ritchin,.F, Nivat, A., Cavalchini, P., Bossan,.E, Shore, M., Paglen, T., Broomberg, A., … Linfield, S. (2011). The anxiety of images. *Aperture, 204*(Fall 2011), 50–73. https://www.jstor.org/stable/24473667

Sullivan, C. (2010). *Art practice as research, inquiry in the visual arts* (2nd ed.). SAGE Publications Inc.

Taggart, E., & Cole, M. (2022). The history of Camera Obscura and how it was used as a tool to create art in perfect perspective. https://mymodernmet.com/camera-obscura/. Accessed June 27, 2024.

Tom, D., & Kearns, A. (n.d.). Learn the secret to great smartphotography. https://www.Adobe.com/creativecloud/photography/discover/smartphone-photgraphy.html. Accessed March 12, 2025.

Trenholm, R. (2021). History of digital cameras: From '70s prototypes to iPhone and Galaxy's everyday wonders. The digital camera has come a long way. https://www.cnet.com/tech/computing/history-of-digital-cameras-from-70s-prototypes-to-iphone-and-galaxys-everyday-wonders/. August 12, 2024.

Tsakiridou, C. A. (2021). Icons and photographs: An ontological and aesthetic comparison. *Analogia, the Pemptousia Journal for Theological Studies, 14*, 63–80.

United Nations. (2015). Sustainable consumption and production. https://sdgs.un.org/topics/sustainable-consumption-and-production. Accessed July 14, 2024

United Nations Brundtland Commission. (1987). Our common future: Report of the world commission on environment and development. *United Nations Brundtland Commission.* http://www.un-documents.net/our-common-future.pdf.

United Nations Environment Programme. (2015). Sustainable consumption and production indicators for the future SDGS. March 2015 UNEP Discussion Paper. https://sdgs.un.org/sites/default/files/publications/2301SCP%20indicators.pdf. Accessed July 14, 2024.

Van Wyk, A. (2024). Veldskool and the loss of innocence. *Vrye Weekblad*. https://www.vryeweekblad.com/en/opinions-and-debate/2024-05-10-veldskool-and-the-loss-of-innocence/. Accessed January 16, 2025.

Chapter 12
The Role of Internationalisation on Research Impact

Kaitano Dube

Abstract Internationalisation is a critical component of university life and development. While internationalisation craft has been mastered by major universities which have used it to exert their dominance and ensure their appearance in university rankings and attract top students and staff nationally and internationally. For many universities, the role and impact of internationalisation are poorly understood. This study examines the impact of internationalisation on research metrics. This study used SciVal data which depend on Scopus-indexed publications. The study looked at 5 universities of technology for the period between 2015 and 2024. The study revealed that universities with strong internationalisation have better research output. Publications co-authored with international collaborators enjoyed better citations which surpassed the global field-weighted cited-impact performance. This far outweighs institutional and national co-authorship. The study also noted a significant decline in the number of countries which cite research by South African researchers, which could be a sign of a decline in the quality of research or a decline in South African research influence. This calls for a rethink by higher education to address challenges faced by researchers, including declining funding and making it easier for international researchers to study and work in South Africa. Fostering greater international collaboration can go a long way in addressing the decline in research output and impact.

Keywords Xenophobia in academia · SDG 4 · International collaboration · South African higher education · Higher education reform · University ranking

---

K. Dube (✉)
College of Economic and Management Sciences , University of South Africa , Pretoria , South Africa
e-mail: dubekaitano@gmail.com

Emirates Aviation University, Dubai Academic City, Dubai, UAE

K. Dube and L. M. Tjabane (eds.), *Advancing Sustainable Development Goals in Higher Education*, https://doi.org/10.1007/978-3-032-27646-9_12

12.1 Introduction

Internationalisation is a critical pillar of development for many states and countries. Internationalisation is crucial for all universities worldwide, with several benefits accruing to them. Universities worldwide view internationalisation as a catalyst for several aspects. Among these is the fact that universities utilise internationalisation to foster and enhance their global presence and standing. In some respects, internationalisation has also been seen as a catalyst for enhancing education quality in line with the dictates of SDG 4 on quality education. de Wit and Altbach (2021) argues that internationalisation is a critical aspect of university reform agenda which has evolved over time.

According to Rajkhowa (2024), the benefits of internationalisation become apparent when universities adopt a whole-institution approach, where there is clear intent, investment, and infrastructure alignment which aligns strategy, policy, and day-to-day practice in all spheres. Consequently, university internationalisation is touted as fostering intercultural (Gao & Liu, 2023) into the university system to improve the university's offerings. Internationalisation allows universities to integrate international, intercultural, and global dimensions into their policies and practices. This approach allows universities to balance their need for local and international relevance. In many ways, universities with an international outlook are able to attract international staff and students (Tight, 2022), which improves the feel of the university, enriches the student experience, and helps produce students who have international competencies and appeal. Tran et al. (2023) noted that university internationalisation has been used by developed countries in some instances to address skills gap and foster national development.

Internationalisation also allows universities to benchmark skills, ensuring the production of an internationally relevant and globally competitive workforce (Fasanmi, 2023). The approach also allows the integration of local, indigenous, and international knowledge and expertise (Maringe, 2009). Similarly, the cross-pollination of local and international staff offers (Chankseliani, 2022) numerous advantages, including the ability to collaborate on research and innovation in a manner that also enhances the curriculum. Hughes (2018) noted that universities' international engagement is also beneficial as it fosters economic development.

Regardless of the numerous benefits of internationalisation, this branch is often overlooked and has recently faced challenges due to increased exaggerated nationalism, anti-foreign sentiments, and xenophobia. These sad developments threaten to undo some of the gains and the spirit of academic harmony and collaboration. Several countries have revised their legislation to make it difficult for international staff to move. The increased visa restrictions in some countries have also proved to be a big hurdle for developing countries, which cannot be trusted with easy access to developed countries with visa restrictions imposed by the USA being a case in point. For developing countries, visa restrictions add to the existing challenges of financial constraints which complicate the internationalisation agenda for developing countries.

In a fast-paced changing world, researchers are seen to be some of the main beneficiaries of internationalisation projects have been researchers and innovators (Elkin et al., 2008). While the benefits of internationalisation are documented in some geographic regions, the benefits of internationalisation in developing country universities have not been fully understood. To this end, this study examines the benefits of internationalisation on research impact. Research impact is one of the tools used to rank universities and researchers. Insights into how internationalisation are needed to justify the huge investments that are often made in such ventures. Collaboration between the Global North and Global South is often seen as largely skewed in favour of the rich elite universities in the developed world.

12.2 Research Methodology

The study relies on data obtained from Scival Scopus for Central University of Technology, Tshwane University of Technology, Durban University of Technology, Cape Peninsula University of Technology, and Vaal University of Technology Scopus-indexed publications for the period 2015 to 2024. SciVal allows researchers to visualise research performance and offers an opportunity for benchmarking by institutions relative to peers, allowing them to forge research partnerships and analyse research trends. SciVal is updated every week and has analytical tools that can be customised to allow for various types of analysis. With respect to the current studies, an institutional analysis was conducted for the selected 5 universities of technology in South Africa. SciVal's powerful analytical capabilities were utilised to analyse the data for each university, and data analysis was conducted in August 2025.

12.3 Results and Discussions

The study found that internationalisation significantly impacts the transformation of research. Figure 12.1 shows the results for the five selected universities of technology (UoT) and the analysed data. The data revealed that universities of technology in South Africa have various levels of international collaboration, with the Central University of Technology collaborating on at least 22.3% of publications, Durban University of Technology 44.8%, Cape Peninsula University of Technology 32.7%, Tshwane University of Technology 44.8% and Vaal University of Technology 26.8%. The remaining percentage for each university was either national or institutional collaboration. In the selected UoTs, the institutions have varying levels of research productivity. The impact of collaboration on research output shows that universities with stronger international collaboration had the highest number of research outputs compared to those with weaker collaboration.

International, national and institutional collaboration by at the Central University of Technology in the selected year range.

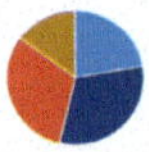

Metric		Scholarly Output	Citations	Citations per Publication	Field-Weighted Citation Impact
International collaboration	22.3%	435	8,020	18.4	1.15
Only national collaboration	31.5%	614	7,782	12.7	0.84
Only institutional collaboration	31.0%	605	3,829	6.3	0.70
Single authorship (no collaboration)	15.1%	295	2,473	8.4	0.68

International, national and institutional collaboration by at the Durban University of Technology in the selected year range.

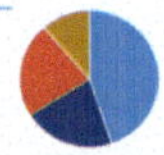

Metric		Scholarly Output	Citations	Citations per Publication	Field-Weighted Citation Impact
International collaboration	44.8%	2,648	107,815	40.7	2.91
Only national collaboration	21.4%	1,261	14,155	11.2	0.97
Only institutional collaboration	22.4%	1,325	17,782	13.4	0.94
Single authorship (no collaboration)	11.4%	673	2,793	4.2	0.54

International, national and institutional collaboration by at the Cape Peninsula University of Technology in the selected year range.

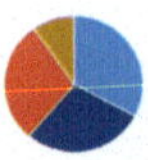

Metric		Scholarly Output	Citations	Citations per Publication	Field-Weighted Citation Impact
International collaboration	32.7%	1,467	40,469	27.6	1.85
Only national collaboration	28.7%	1,288	13,674	10.6	0.81
Only institutional collaboration	28.7%	1,288	11,933	9.3	0.76
Single authorship (no collaboration)	9.9%	445	2,675	6.0	0.57

International, national and institutional collaboration by at the Tshwane University of Technology in the selected year range.

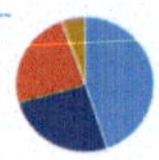

Metric		Scholarly Output	Citations	Citations per Publication	Field-Weighted Citation Impact
International collaboration	44.8%	3,363	75,745	22.5	1.39
Only national collaboration	26.0%	1,949	23,604	12.1	0.85
Only institutional collaboration	23.6%	1,767	15,887	9.0	0.69
Single authorship (no collaboration)	5.6%	419	2,973	7.1	0.66

International, national and institutional collaboration by at the Vaal University of Technology in the selected year range.

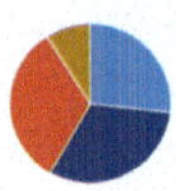

Metric		Scholarly Output	Citations	Citations per Publication	Field-Weighted Citation Impact
International collaboration	26.8%	511	7,621	14.9	1.00
Only national collaboration	31.8%	607	8,065	13.3	0.97
Only institutional collaboration	32.4%	618	5,887	9.5	0.83
Single authorship (no collaboration)	9.1%	173	981	5.7	0.62

Fig. 12.1 Research collaboration and its impact on citation for universities of technology

Figure 12.2 shows the trend of citing countries for technology universities. This data revealed that there was an exponential growth in the number of citing countries for research conducted by UoTs, with the most significant number of citations received between 2016 and 2022. However, there is a decline in the number of countries citing research conducted by UoTs. This could be a sign of a decline in research quality and influence by South African scholars. This trend comes at a time when there has been a persistent decline in funding for higher education institutions which could have adversely affected research activities and quality of the conducted research. It is equally unclear how the country's extreme nationalism and xenophobia could be an attributing factor to this trend. The country has also been suffering bouts of xenophobic attacks, largely targeting black academics from African countries. Consequently, there has been a decline in the number of international students in

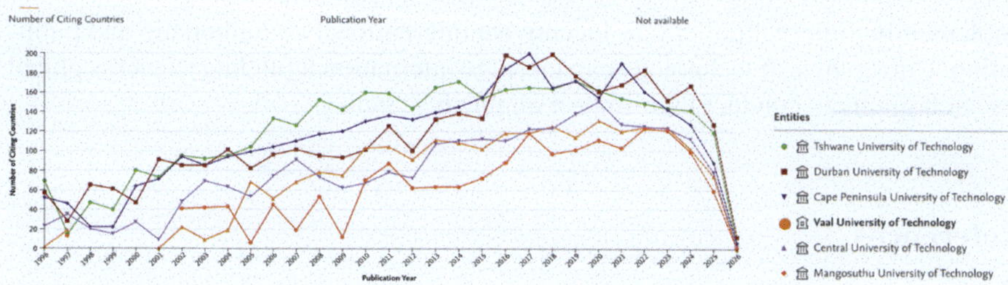

Fig. 12.2 Number of research citing countries for universities of technologies

South African universities (Quinlan & Singh, 2025). The hostile attitude towards international staff could also be a factor in the declining citing countries. These could be factors affecting the country's research impact and influence.

Figures 12.1 and 12.2 show the significant benefits of internationalisation. Evidence shows that internationalisation results in improved citation impacts of publications that are co-authored with international authors when compared to publications with either institutional or international co-authorship. The field-weighted citations for each institution are clearly much higher for international co-authored papers. The institutions that have benefitted the most from this international collaboration are Durban University of Technology which has 2.91(FWCI), Cape Peninsula University of Technology 1.85 (FWCI), Tshwane University of Technology 1.39 (FWCI), Central University of Technology 1.15 (FWCI), and Vaal University of Technology 1(FWCI), with all universities being at par with international performance (VUT), while all the other universities outperform international performance. The picture is different when one considers institutional and national performance, where every institution performs below 1 FWCI.

12.4 Conclusion and Recommendations

This study examined the impact of internationalisation on research citation impact. A study which utilised data from SciVal for five universities of technology found that research collaboration with international authors and institutions was beneficial for the impact of research output by South African academics. Research conducted with international researchers outperformed the citations of those studies co-authored with national and institutional collaborators. While internationalisation is beneficial to international universities, evidence points to a decline in the number of countries citing South African work.

To sustain and improve the research impact of South African universities, this study proposes several interventions, including a call for universities to adopt a whole-institution approach which ensures alignment of strategy, policy, and infrastructure to support internationalisation by university institutions. There is a need to prioritise international research partnerships by availing adequate institutional funding and implementing internal policies to incentivise international co-authorship and publication. Universities of technology can leverage international funding to access global research grants to conduct high-impact citable research.

References

Chankseliani, M. (2022). Internationalisation of universities. In M. Chankseliani (Ed.), *What happened to the Soviet University?* Oxford University Press. https://doi.org/10.1093/oso/9780192849847.003.0004

de Wit, H., & Altbach, P. G. (2021). Internationalization in higher education: Global trends and recommendations for its future. *Policy Reviews in Higher Education, 5*(1), 28–46. https://doi.org/10.1080/23322969.2020.1820898

Elkin, G., Farnsworth, J., & Templer, A. (2008). Strategy and the internationalisation of universities. *International Journal of Educational Management, 22*(3), 239–250. https://doi.org/10.1108/09513540810861874

Fasanmi, S. A. (2023). Internationalisation of higher education in the new normal in universities: University management perspectives. *EUREKA: Social and Humanities*, (4), 27–36. https://doi.org/10.21303/2504-5571.2023.002988

Gao, Y., & Liu, J. (2023). Innovating policies for university internationalisation in the changing post-pandemic global field. *Policy Reviews in Higher Education, 7*(1), 78–97. https://doi.org/10.1080/23322969.2022.2114531

Hughes, R. (2018). University internationalisation strategies and public engagement. In J. James, J. Preece, & R. Valdés-Cotera (Eds), *Entrepreneurial learning city regions: Delivering on the UNESCO 2013, Beijing declaration on building learning cities* (pp. 35–57). Springer International Publishing. https://doi.org/10.1007/978-3-319-61130-3_3

Maringe, F. (2009). Strategies and challenges of internationalisation in HE: An exploratory study of UK universities. *International Journal of Educational Management, 23*(7), 553–563. https://doi.org/10.1108/09513540910990799

Quinlan, O., & Singh, T. (2025). *Latest news—How can internationalisation regain momentum in South Africa?* https://www.ru.ac.za/perspective/latestnews/how_can_internationalisation_regain_momentum_in_south_africa.html

Rajkhowa, G. (2024). Internationalisation for all: Rethinking university internationalisation. *Space and Culture, India, 11*(4), 5–19. (Worldwide). https://doi.org/10.20896/saci.v11i4.1341

Tight, M. (2022). Internationalisation of higher education beyond the West: Challenges and opportunities—The research evidence. *Educational Research and Evaluation, 27*(3–4), 239–259. https://doi.org/10.1080/13803611.2022.2041853

Tran, L. T., Jung, J., Unangst, L., & Marshall, S. (2023). New developments in internationalisation of higher education. *Higher Education Research & Development, 42*(5), 1033–1041. https://doi.org/10.1080/07294360.2023.2216062

Part VI
Conclusion and Recommendations

Chapter 13
Innovative Approaches to Teaching, Learning, and Student Success in South Africa, an SDG Perspective

Kaitano Dube, Masebala Liemiseng Tjabane, and Muntuwenkosi Chili

Abstract This chapter serves as the closing chapter for this edited book. The book reflects on how innovative teaching, learning, and research practices in South Africa's higher education sector can advance the Sustainable Development Goals, with an emphasis on SDG4 and student success. The edited book is set against the backdrop of post COVID-19 disruption, rapid technological advances including AI and cyber-attacks, and increased global geopolitical tensions. This chapter demonstrates how higher education institutions have been forced to rethink pedagogy, curriculum design, assessment, and institutional support structures. Deriving evidence from previous chapters, this chapter shows that student success is shaped by interconnected factors such as the lecturer, the student, the institution, and other role players which need to be properly synchronised in a holistic manner to ensure sustainability. This chapter reinforces the role of experiential, problem-based, immersive, and technology-enhanced learning in fostering a quality education system which delivers sustainability.

Keywords Education sustainability · Quality education · Student success · SDGs higher education · Higher education policy

K. Dube (✉)
College of Economic and Management Sciences, University of South Africa, Pretoria, South Africa
e-mail: dubekaitano@gmail.com

Emirates Aviation University, Dubai Academic City, Dubai, UAE

M. L. Tjabane · M. Chili
Centre of Academic Development, Vaal University of Technology, Vanderbijlpark, South Africa
e-mail: masebala@vut.ac.za

K. Dube and L. M. Tjabane (eds.), *Advancing Sustainable Development Goals in Higher Education*, https://doi.org/10.1007/978-3-032-27646-9_13

13.1 Background and Introduction

Global changes over the past decades have forced a rethink of teaching and learning, with several emerging trends aimed at ensuring system sustainability. The higher education system has undergone significant transformation owing to several internal and external factors. Changes in global geopolitics and local political factors have had a huge impact on some higher education institutions across the world.

One factor that has impacted teaching and learning with implications for student success has been the advent of the COVID-19 pandemic in 2020. The COVID-19 pandemic was disruptive to teaching and learning norms in many respects (Harmey & Moss, 2023; Parveen et al., 2022; Uwizeyimana, 2022), and the impact of the pandemic will be difficult to reconfigure or reverse in the near and medium future. The COVID-19 pandemic altered several aspects of teaching and learning as it transformed the socio-economic conditions which are critical support structures for students and higher education facilitators.

The trauma and fear triggered by the COVID-19 pandemic forced physical and social distancing (Nhamo et al., 2020) with disruptive consequences on curriculum delivery, institutional practice, and transformed how people in higher education behave. Several institutions changed their curriculum delivery and student assessment methods. The ramifications of this reset are still being felt in higher education institutions and corporations across the world.

The impact of the COVID-19 pandemic was followed by the rapid rollout of AI through the advent of ChatGPT and other chatbots. These technological developments have had a significant impact on teaching, learning, and assessment (Dube et al., 2025). This further complicated matters for a system that was battling to recover from the impact of the COVID-19 pandemic itself. The post-pandemic period was characterised by fatigued and traumatised learning communities (Oberg et al., 2023). This called for a rethinking of the kind of education system which had to emerge from the ashes of the pandemic. The South African education system has also faced similar challenges, if not more complex challenges, with the pandemic emerging soon after the chaotic fees must fall student' uprisings (Dube, 2024). These challenges call for innovative approaches to both teaching and learning. With the world fast pacing to 2030, such innovations must encapsulate the global goals famously known as the Agenda for Sustainable Development. The SDGs, which are all-inclusive, will lapse in 2030.

13.2 Teaching and Learning Success Drivers

Student teaching is often affected by several factors (Malandrakis et al., 2021), including those which affect students, institutions, and learning facilitators. Facilitator-related factors include the subject knowledge capacity of the facilitator. Facilitators who are knowledgeable about the subject matter stand a better chance of delivering content compared to those who may not be as knowledgeable about the subject matter. The facilitators are skilled and approachable, with discipline skills which promote good behaviour (Gibbs & Coffey, 2004). Facilitators with good mental and psychological well-being are critical to delivering quality education. One of the biggest challenges faced by higher education post-pandemic has been the lack of facilitators with such capacity. Paying attention to some of these effects is critical in the delivery of quality education which ensures students' success.

There are equally student-related factors, which affect the quality of teaching, learning, and success factors (Moriña & Biagiotti, 2022; Soleimani et al., 2024; Yuebo et al., 2024). The well-being of students, including their physical and mental health, is critical to ensuring a healthy education ecosystem. Psychosocial factors have become critical considerations for Higher Education Institutions (HEI) after the pandemic. Most students experienced a serious knock from the strain of the pandemic, with severe ramifications for their psychosocial well-being. The learner's potential is also a critical consideration, as it is shaped by various factors such as aptitude, attitude, and innate ability. Aspiration, life goals, and readiness are equally critical factors which act as success drivers for university students.

Other factors that have been found to affect teaching and learning are environmental factors, which range from support factors such as the infrastructure of higher education institutions (Chaudhary & Singh, 2022) which affects the delivery of learning content to students. The availability of learner material, resourcing of libraries, labs, classroom environment, availability of technology, and physical comfort of students and facilitators play a role in teaching and learning success. In many respects, several institutions have also onboarded learner management systems to be a central cog and part and parcel of teaching and learning in the delivery of tuition and to ensure inclusivity.

In the past couple of years, between 2020 and 2025, local universities have also been making frantic efforts to continuously tap into the SDG framework to improve the quality of teaching and learning such as the adoption of approaches which enhance student skills and capabilities such as the adoption of experiential learning, adopting enhanced technology teaching approaches, and ensuring the infusion of an interdisciplinary approach to curriculum development and delivery (Mawonde & Togo, 2019; Nhamo et al., 2024). This approach is aimed at fostering sustainability awareness and promoting action towards sustainability.

Some of the approaches adopted by facilitators include experiential learning (Murrell & Claxton, 1987) and immersive learning. The approach encompasses various techniques, such as the use of simulation games, field-based projects (Fedesco et al., 2020), and roleplays to assist students in grasping sustainability concepts which often call for trade-offs to be made, especially in resolving potential conflicts of interest.

Institutions have also been developing immersive programs (Buragohain et al., 2024; Sviridova et al., 2023) to cultivate cross-cultural competencies and develop critical leadership skills by placing SDGs learning in real-world scenarios of daily living. The use of case studies on sustainability in some courses has raised awareness and fostered pro-environmental behaviour. This approach, for example, has been utilised in hospitality and tourism courses.

Problem-based learning (Yew & Goh, 2016) has also been utilised to foster and promote quality teaching and learning. The approach includes utilising students in the co-designing of campus sustainability initiatives with a view to enhancing students' problem-solving skills and ensuring workplace readiness. The projects also prompt students to engage in critical thinking and develop critical thinking skills and system perspectives to allow them to address the real complex challenges faced by the world.

As mentioned earlier, some institutions are embracing technology-enhanced education by utilising various digital strategies in teaching and learning, including the use of VR, gamification, and AI to expand access to personalised learning (Chere & Wayi-Mgwebi, 2024). It is also used to foster global collaboration. In many ways, institutions are equally embracing context-based approaches. Some courses are also embracing ecopedagogy in environmental courses, fostering critical thinking by linking topics studied by students to SDGs frameworks.

In many ways, the SDGs have also found resonance in the research space, with lecturers participating in the SDG. Some institutions have performed far better than others, with various institutions focussing on various SDG themes. Research is a critical component of teaching, with many higher education institutions relying on research-driven teaching and learning to produce quality and competent graduate. The following section examines how some universities of technology have aligned research with teaching and learning.

13.3 SDG Aligned Research in Universities of Technology

Universities of technology have been doing their bit to ensure that research is aligned with the SDGs to ensure relevance and alignment. Figure 13.1 demonstrates the various SDGs that the universities of technology in South Africa are focussing on and their performance. Various universities have different focusses on SDGs. The Vaal University of Technology, Tshwane University of Technology, Durban University of Technology, and Cape Peninsula University of Technology have a strong focus on water and sanitation. The Vaal and Durban universities of technology have a particularly strong lead in this endeavour. This can be understood in the context of the pressing water and sanitation challenges in the country resulting from poor management and governance, climate change, and water pollution in South Africa which affect freshwater and marine water resources (Chikodzi et al., 2021; Mdoda et al., 2024).

South Africa has several water challenges which require attention and innovative solutions to address the adverse impacts of water shortages and pollution. Focussing on this as part of the curriculum can be beneficial in many respects. Another SDG that is popular with researchers in Universities of Technology (UoTs) is SDG 12, which focusses on responsible and sustainable consumption. The Vaal University of Technology (VUT) has given considerable attention to this area. So is Durban University of

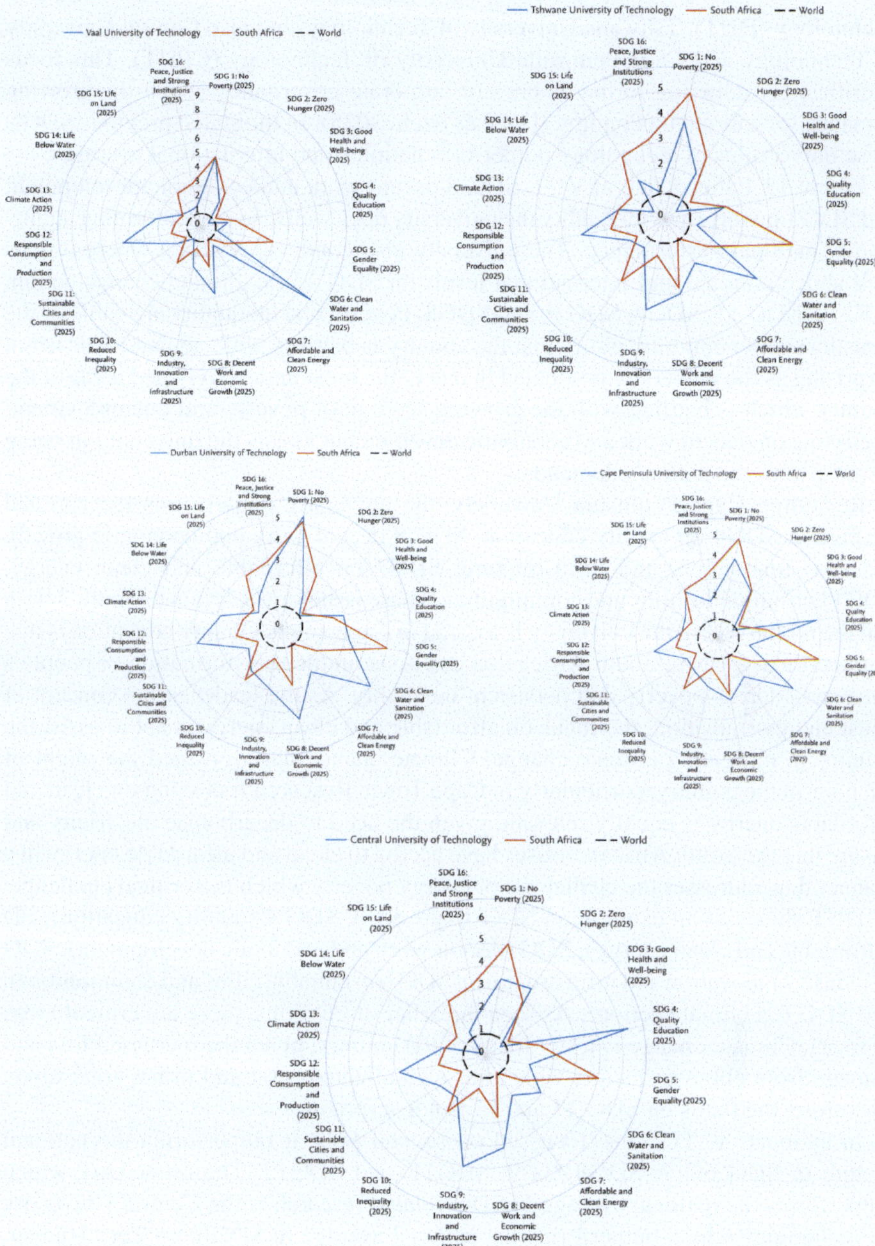

Fig. 13.1 SDGs focus areas for universities of technology in South Africa

Technology (DUT), Tshwane University of Technology, and also Central University of Technology and Cape Peninsula University of Technology (CPUT). This focus is critical for adhering to the dictates of corporate governance, given the growing demands for a circular economy (Dube & Ezeh, 2025). In the stated previous SDGs, these universities of technology outperform national and international averages.

Figure 13.1 shows that at VUT, SDG 8 decent work and economic development and SDG 1 poverty and inequality the university does well also outperforming significantly national performance. The university also conducts research where it does well above national and international levels on SDG 13 on climate change action, SDG 12, SDG 11, SDG8, SDG 9, and SDG8. Poverty and inequality are some of the pressing socio-economic issues that the country is battling with, so that focus is not surprising as the university is situated in one of the most underprivileged areas of the country which is battling with the growing scourge of poverty and unemployment. Focussing on decent work and economic development assists the university in being responsive to the nation's demands.

Regarding Cape Peninsula University, the university outperforms the national average in SDG4 on quality education, SDG 8 decent work and economic growth, Industry, Innovation, and infrastructure, SDG7 on affordable and clean energy, SDG11 sustainable cities and communities. These issues are relevant to South Africa and the region where the university is located in Cape Town. Quality education is one of the critical areas in South Africa that can assist in the transformation of people's lives and address poverty and persistent inequality. As the leader in the concept of a just energy transition, the focus on affordable and clean energy is apt to assist the country in mitigating climate change. Climate change has worsened the plight of the poor in the country, particularly in Cape Town. Research focussing on clean and affordable energy is equally relevant, given the need to decarbonise electricity and ensure that the South African masses have access to clean and affordable energy in a manner that addresses the challenges of energy poverty which is a critical challenge.

DUT shines as an example in SDG 1 (poverty), SDG 4 (quality education), SD (affordable and clean energy), SDG (decent work and economic development), SDG (9 industry innovation and infrastructure), SDG (sustainable cities and communities), and SDG (on climate action). As seen in earlier discussions, these are critical areas of focus. Climate change is a key focus of the country, given the increased loss and damage from climate change, which has also affected rural and urban areas alike. Therefore, the focus on SDG 13 and 11 is not surprising.

In as much as TUT has focussed on several SGs, it outperforms the national picture in SDG 6G 7, SDG8 8G 9, SDG11, and SDG12. The other UoT which outperforms the national average in SDG-related research is the Central University of Technology which outperforms the national average in SDG 2 on Zero Hunger, SDG4 on quality education, SDG 7 on Affordable and clean energy, and SDG 8, 9, and 12 on responsible consumption and production.

This edited book delves into cross-cutting issues that have implications for higher education institutions going forward and reaches the following conclusions and recommendations.

13.4 Conclusions and Policy Recommendations

This edited book examines how South African higher education systems are developing innovative approaches to ensure the achievement of quality education in line with SDG4 and assist in the creation of a better world. The book can make the following policy recommendations that emerge from various chapters of this book to assist in the improvement of education and higher education system in South Africa.

The study recommends that higher education in South Africa should assist in professional development programmes, such as the TAU Fellowship, to produce competent facilitators to produce SDG 4 ready candidates. Given the geographical, social, and political realities of South Africa, quality education and student success can be enhanced using multidimensional early warning systems which will ensure that proactive support is afforded to students which can increase success and throughput rates.

The decolonisation of the curriculum is a critical component of ensuring the delivery of a quality and progressive curriculum. To yield the desired results, there is also a need for the government and policymakers to push for funding for hybrid infrastructure which ensures the availability of high-speed connectivity and ensures digital competency and access to education for all citizens in the spirit of lifelong learning.

In addition, higher education institutions need to develop tailored support for extended curriculum programmes to ensure inclusivity for the urban and rural populace. Campuses also have to make strides to ensure that they are not mere theorists of sustainability but have to make strides to ensure that they actively participate in sustainability initiatives such as embarking and adopting green technologies and fostering a culture of sustainability.

References

Buragohain, D., Deng, C., Sharma, A., & Chaudhary, S. (2024). The impact of immersive learning on teacher effectiveness: A systematic study. *IEEE Access, 12*, 35924–35933. https://doi.org/10.1109/ACCESS.2024.3373541

Chaudhary, P., & Singh, R. K. (2022). A meta analysis of factors affecting teaching and student learning in higher education. *Frontiers in Education, 6*. https://doi.org/10.3389/feduc.2021.824504

Chere, M., & Wayi-Mgwebi, N. (2024). Integration of generative artificial intelligence in higher education: Pedagogy factors and best practices. In K. Dube (Ed.), *Redefining education and development: Innovative approaches in the era of the sustainable development goals* (pp. 93–112). Springer Nature Switzerland. https://doi.org/10.1007/978-3-031-69954-2_7

Chikodzi, D., Dube, K., & Ngcobo, N. (2021). Rethinking harbours, beaches and urban estuaries waste management under climate-induced floods in South Africa. In G. Nhamo & L. Chapungu (Eds), *The increasing risk of floods and tornadoes in Southern Africa* (pp. 127–140). Springer International Publishing. https://doi.org/10.1007/978-3-030-74192-1_7

Dube, K. (2024). Challenges and opportunities facing higher education in South Africa. In K. Dube (Ed.), *Redefining education and development: Innovative approaches in the era of the sustainable development goals* (pp. 11–27). Springer Nature Switzerland. https://doi.org/10.1007/978-3-031-69954-2_2

Dube, K., & Ezeh, P. C. (2025). A global perspective on ecotourism marketing trends: A review. *Sustainability, 17*(13), 6035. https://doi.org/10.3390/su17136035

Dube, K., Gwatidzo, S., & Maleho, L. (2025). Opportunities and challenges of AI adoption in higher education, a post COVID-19 analysis. In *2025 3rd international conference on business analytics for technology and security (ICBATS)*, pp. 1–5. https://doi.org/10.1109/ICBATS66542.2025.11258206

Fedesco, H. N., Cavin, D., & Henares, R. (2020). Field-based learning in higher education: Exploring the benefits and possibilities. *Journal of the Scholarship of Teaching and Learning, 20*(1). https://doi.org/10.14434/josotl.v20i1.24877

Gibbs, G., & Coffey, M. (2004). The impact of training of university teachers on their teaching skills, their approach to teaching and the approach to learning of their students. *Active Learning in Higher Education, 5*(1), 87–100. https://doi.org/10.1177/1469787404040463

Harmey, S., & Moss, G. (2023). Learning disruption or learning loss: Using evidence from unplanned closures to inform returning to school after COVID-19. *Educational Review, 75*(4), 637–656. https://doi.org/10.1080/00131911.2021.1966389

Malandrakis, G., Bara, E.-Z., & Gkitsas, S. (2021). Perceived factors affecting student-teachers' learning about social urban sustainability. *Environmental Education Research, 27*(9), 1360–1382. https://doi.org/10.1080/13504622.2021.1927991

Mawonde, A., & Togo, M. (2019). Implementation of SDGs at the University of South Africa. *International Journal of Sustainability in Higher Education, 20*(5), 932–950. https://doi.org/10.1108/IJSHE-04-2019-0156

Mdoda, S. S., Dube, K., & Montsiemang, T. (2024). Tackling water and waste management challenges within the tourism and hospitality industry: A sustainable development goals perspective. *Water, 16*(24), 3545. https://doi.org/10.3390/w16243545

Moriña, A., & Biagiotti, G. (2022). Academic success factors in university students with disabilities: A systematic review. *European Journal of Special Needs Education, 37*(5), 729–746. https://doi.org/10.1080/08856257.2021.1940007

Murrell, P. H., & Claxton, C. S. (1987). Experiential learning theory as a guide for effective teaching. *Counselor Education and Supervision, 27*(1), 4–14. https://doi.org/10.1002/j.1556-6978.1987.tb00735.x

Nhamo, G., Chapungu, L., & Dube, K. (2024). Universities and sustainable development goals localisation: Insights from academic staff in Zimbabwe. *Cogent Education, 11*(1), 2428877. https://doi.org/10.1080/2331186X.2024.2428877

Nhamo, G., Dube, K., & Chikodzi, D. (2020). *Counting the cost of COVID-19 on the global tourism industry*. Springer International Publishing. https://doi.org/10.1007/978-3-030-56231-1

Oberg, G., Carroll, A., & Macmahon, S. (2023). Compassion fatigue and secondary traumatic stress in teachers: How they contribute to burnout and how they are related to trauma-awareness. *Frontiers in Education, 8*. https://doi.org/10.3389/feduc.2023.1128618

Parveen, K., Tran, P. Q. B., Alghamdi, A. A., Namaziandost, E., Aslam, S., & Xiaowei, T. (2022). Identifying the leadership challenges of k-12 public schools during COVID-19 disruption: A systematic literature review. *Frontiers in Psychology*. https://doi.org/10.3389/fpsyg.2022.875646

Soleimani, S. M., Jaeger, M., Faheiman, A., & Alaqqad, A. R. (2024). Success factors of recently implemented eLearning methods at higher education institutions in Kuwait. *Quality in Higher Education, 30*(1), 130–142. https://doi.org/10.1080/13538322.2022.2132702

Sviridova, E., Yastrebova, E., Bakirova, G., & Rebrina, F. (2023). Immersive technologies as an innovative tool to increase academic success and motivation in higher education. *Frontiers in Education, 8*. https://doi.org/10.3389/feduc.2023.1192760

Uwizeyimana, D. E. (2022). Analysing the importance of e-government in times of disruption: The case of public education in Rwanda during Covid-19 lockdown. *Evaluation and Program Planning, 91*, Article 102064. https://doi.org/10.1016/j.evalprogplan.2022.102064

Yew, E. H. J., & Goh, K. (2016). Problem-based learning: An overview of its process and impact on learning. *Health Professions Education, 2*(2), 75–79. https://doi.org/10.1016/j.hpe.2016.01.004
Yuebo, L., Halili, S. H., & Razak, R. A. (2024). Online learning success model for adults in open and distance education in Western China. *PLoS ONE, 19*(2), Article e0297515. https://doi.org/10.1371/journal.pone.0297515

Index

K. Dube and L. M. Tjabane (eds.), *Advancing Sustainable Development Goals in Higher Education*, https://doi.org/10.1007/978-3-032-27646-9

The manufacturer's authorised representative in the EU is Springer Nature Customer Service Centre GmbH, Europaplatz 3, 69115 Heidelberg, Germany. If you have any concerns regarding our products, please contact ProductSafety@springernature.com

Printed and bound by CPI Group (UK) Ltd, Croydon, CR0 4YY
21/09/2026
02219985-0005